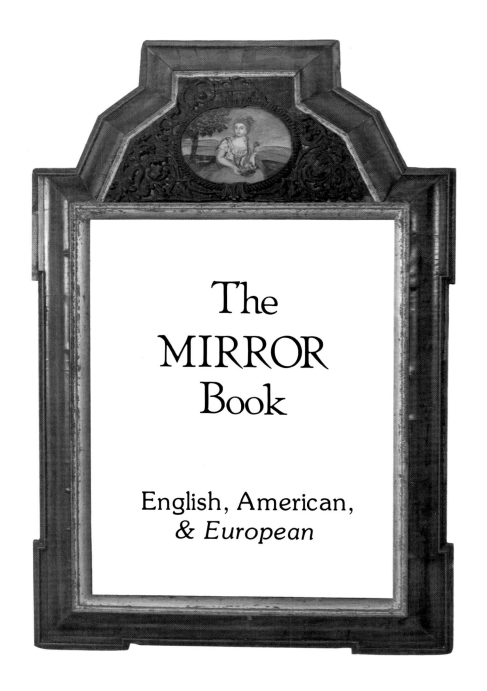

The MIRROR Book

English, American, & European

by
Herbert F.Schiffer

Box E, Exton, Pennsylvania 19341

Other books by the author

Woods We Live With, with Nancy Schiffer

Shaker Architecture

Chinese Export Porcelain, Standard Patterns and Forms, 1780 to 1880, with Peter and Nancy Schiffer

China For America, Export porcelain of the 18th and 19th centuries, with Peter and Nancy Schiffer

The Brass Book, American, English and European, 15th century to 1850, with Peter and Nancy Schiffer

Antique Iron, English and American, 15th century through 1850, with Peter and Nancy Schiffer

Early Pennsylvania Hardware

Printed in the United States of America

ISBN: 0-916838-82-X

**Schiffer Publishing Limited
Box E
Exton, Pennsylvania 19341**

Jacket: Constitution or architectural mirror, walnut veneer and gilded wood, English, circa 1740, see #216.
Title page: Courting mirror, walnut veneer, pressed brass, gilded fillet, and eglomise, European, eighteenth century, see #111.

TABLE OF CONTENTS

| Queen Anne and William and Mary | Chippendale | Georgian | Adam | Victorian | Modern |

Plate I.

Plate I. This wooden board has been painted with colored sizing (often called red French clay) matching those used under gilding on mirrors at different periods of English design. The tone at the far left is found under gilding on Queen Anne and William and Mary period mirrors, circa 1680 to 1720. The next tone is found under Chippendale and Rococo gilding on mirrors of the period circa 1740 to 1760. The dark tone next may be found under gilding of Georgian period mirrors. The pale tone next is found under Adam period gilding on mirrors dating circa 1780 to 1810. The next block is shown with two tones which can be found on gilded mirrors of the Empire and Victorian periods, circa 1810 to 1900. The yellow on the right sometimes is found under gilding of the Victorian to Modern periods. Oil gilding was most likely used over the grey and yellow sizing in the late period, while water gilding was used at the earlier periods.

Plate II.

Plate III.

Plate IV.

Plate V.

Plate II. Queen Anne mirror, tortoise shell and stump work embroidery, English, late seventeenth century, 24½″ high, 21 3/8″ wide, see #42. Notice the cross banding on the mirror frame. If this is not cracked and cross banded, the mirror is probably not old. The veneers should be made in small pieces and very thick.

Plate III. Queen Anne mirror, walnut veneer, English, early eighteenth century.

Plate IV. Fret work mirror, mahogany, American, assocaited with James Stokes, Philadelphia, circa 1790, see #394.

Plate V. The back of this mirror shows a variation in color from light at the bottom to gradually darker at the top. As the mirror hung against a wall at a slight angle, closer at the bottom than the top, the greater amount of air, dirt·and soot at the top caused more oxidation and darkening of the wood. Notice the shrinkage of the backboard. If the backboard has not shrunk to show a gap, it is not old. Untouched glue joints, such as at the top of the frame, are a little lighter giving a halo effect. In the eighteenth century many surfaces to be glued were scratched in parallel lines to give a potentially better glue joint. These scratches can be seen on the top crest.

ACKNOWLEDGMENTS

There is no way possible to thank all of the people who took their time to help me by their knowledge and experience. These are some of the people who are impossible to forget. The graciousness, kindness and patience of all is sincerely appreciated. Phyllis Rhodes Abrams, Girard College; Bill Adair; William G. Allman, The White House; Michael Archer, Victoria and Albert Museum; H. Parot Bacot, Anglo-American Museum, Louisiana State University; Simon Bargate, The Hamlyn Group; Clare Baron; Ronald Bauman; George Phillips Beech; Geoffrey de Bellaigue and Julia Harland, Lord Chamberlain's Office; The Bennington Museum; James Billings; Edgar M. Bingham, Junior and Jane D. Crawford, Shreve, Crump & Low, Inc.; Joan Bogart Antiques; Evelyn Brenner, Philadelphia Museum of Art; The British Library; The British Museum; Philip H. Bradley; Alfred Bullard, Inc.; James D. Burke, St. Louis Art Museum; Nancy Carlisle, Winterthur Museum; M. Challen and Anne Buddle, Victoria and Albert Museum; Eleanor S. Clarke; The Colonial Williamsburg Foundation; Clement Conger, The United States Department of State Diplomatic Reception Rooms; The Connecticut Historical Society; John O. Curtis, Old Sturbridge Village; Elizabeth R. Daniel; Bert Denker, Karen Hill, and Mary Piendak, Winterthur D.A.P.I.C. Library; Sanna Deutsch, Honolulu Academy of Arts; Ulysses G. Dietz, The Newark Museum; Gaylord Dillingham; Richard and Eileen Dubrow; William Voss Elder, III, The Baltimore Museum of Art; Jonathan Fairbanks, Museum of Fine Arts, Boston; Amelia N. Farr, Loudoun Mansion; Morris Finkel and Daughter; Samuel T. Freeman and Company; B.T.Galloway, The National Gallery, London; Wendell Garrett, *The Magazine Antiques*; Price Glover; Dr. Sidney M. Goldstein, The Corning Museum of Glass; Carol E. Gordon, Munson-Williams-Proctor Institute; Julia Harland, Assistant to the Surveyor of the Queen's Works of Art; D.S. Hill, Winchester House Property Company, Limited; Cortlandt Y.D. Hubbard; Grace C. Ipock, Tryon Palace Restoration; Ellen Jenkens, Price Glover Incorporated; Carol A. Kim, China Trade Museum; Joe Kindig, III; Dwight Lanmon, The Corning Museum of Glass; Margaret Longmuir, Alnwick Castle; S. Dean Levy; Museum of Early Southern Decorative Arts; Kathryn A. McCutchen, The Fine Arts Committee, United States Department of State; Migs Nutt; Nancy Boyle Press, Baltimore Museum of Art; A. Christian Revi; Franklin Riehlman, The Metropolitan Museum of Art; Albert M. Sack, Israel Sack, Inc.; H. & R. Sandor, Inc.; Margaret B. Schiffer; Science Museum, London; Karol A. Schmiegel, The Henry Francis duPont Winterthur Museum; Esther Schwartz; Mr. and Mrs. Samuel Schwartz; Robert D. Schwarz, Frank S. Schwarze & Son, Philadelphia; Raymond V. Shepherd; R. Innes-Smith, English Life Publications Limited; Myrna Smoot, Los Angeles County Museum of Art; Cyril Staal; David H. Stockwell; Fannie Stokes; Craig and Tarlton, Inc.; William Thomas, gilder, special thanks for his patience with my questions and generosity with his knowledge; Mr. Truman, The Victoria and Albert Museum; Robert T. Trump; Michelin Turpin, a special word of thanks for his unselfish gift of knowledge and fine pictures; The Wallace Collection, London; John S. Walton; John M. Wood, Humberts, Wiltshire; Michael Wright, *Country Life Magazine;* and Patricia Morris Young.

BEGINNINGS

In the literature of antiquity, grave goods, such as weapons, are put into the grave in reverse (mirror image) fashion, corresponding to the concept that the "other world" is a mirror image of our own. (Hartlaub, p. 184)

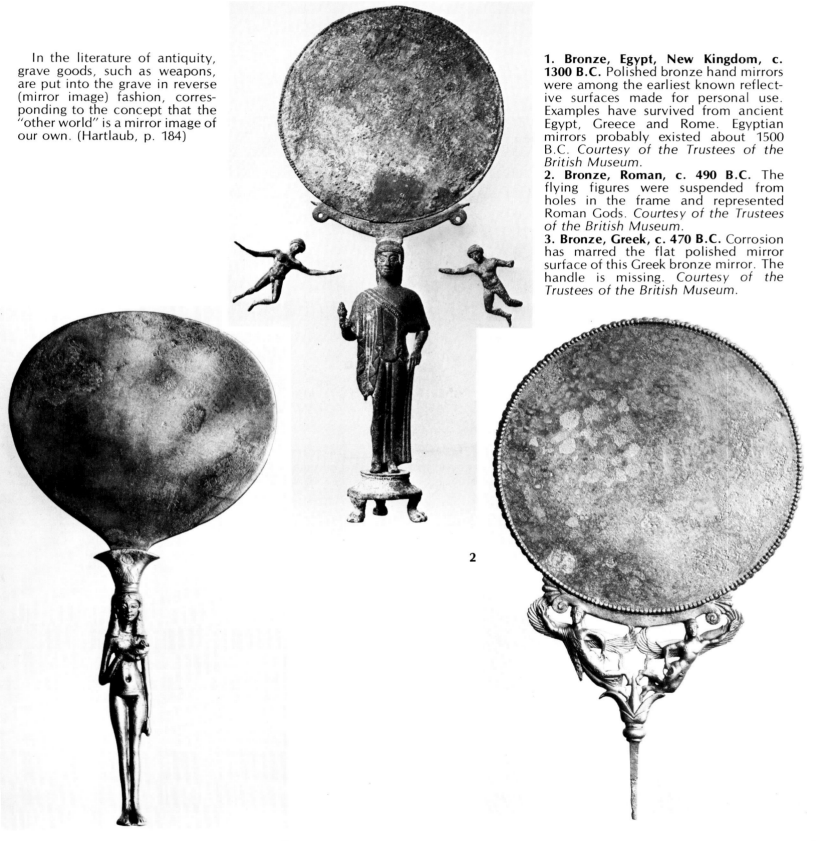

1. Bronze, Egypt, New Kingdom, c. 1300 B.C. Polished bronze hand mirrors were among the earliest known reflective surfaces made for personal use. Examples have survived from ancient Egypt, Greece and Rome. Egyptian mirrors probably existed about 1500 B.C. *Courtesy of the Trustees of the British Museum.*

2. Bronze, Roman, c. 490 B.C. The flying figures were suspended from holes in the frame and represented Roman Gods. *Courtesy of the Trustees of the British Museum.*

3. Bronze, Greek, c. 470 B.C. Corrosion has marred the flat polished mirror surface of this Greek bronze mirror. The handle is missing. *Courtesy of the Trustees of the British Museum.*

2

1

3

8

Plate VI.

Plate VIII.

Plate VII.

Plate IX.

Plate X.

Plate IX. Back of fretwork mirror, mahogany veneer on pine, American, bearing the latest label of John Elliott and Son, Philadelphia, circa 1804 to 1810, 38¾″ high, 19¾″ wide, see #399. This label is the last one used by John Elliott, Junior. It reads: John Elliott & Sons/No. 60, South Front Street between Chestnut and Walnut Streets/Philadelphia/Sell by wholesale and retail,/Drugs and Medicines/Window and Picture glass, of all sizes,/Paints of all kind, dry and ground in oil,/Linseed oil and varnishes/Camel hair pencils & brushes,/Gold & Silver Let ( )/Sand Crucibles,/Dye Woods & c. viz./Logwood & Fustic,/Brazil & Redwood,/Coppera & Alum/also/Looking glasses,/of all sizes, and different patterns,/Old glasses, New Quicksilvered and framed./Country shopkeepers supplied on the best terms, for cash, or/the usual credit." This mirror shows one of the more standard ways of hanging looking glasses. This bent metal plate fit over a hook in the wall.

Plate X. Front of fretwork mirror, Plate IX.

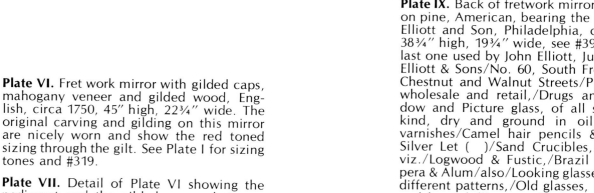

Plate VI. Fret work mirror with gilded caps, mahogany veneer and gilded wood, English, circa 1750, 45″ high, 22¾″ wide. The original carving and gilding on this mirror are nicely worn and show the red toned sizing through the gilt. See Plate I for sizing tones and #319.

Plate VII. Detail of Plate VI showing the pediment and the gilded curves between the basket and side scrolls.

Plate VIII. Detail of Plate VI showing side carved and gilded swag, mahogany veneered frame, and gilded fillet next to the glass.

4

5

6

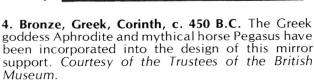

4. Bronze, Greek, Corinth, c. 450 B.C. The Greek goddess Aphrodite and mythical horse Pegasus have been incorporated into the design of this mirror support. *Courtesy of the Trustees of the British Museum.*

5. Bronze, Greek, c. 460 B.C. *Courtesy of the Trustees of the British Museum.*

6. Bronze mirror cover, Etruscan, third to second century B.C. Since polished bronze scratches easily, covers were designed for protection, and these provide a surface for symbolic decoration. The design on the cover of this polished bronze mirror represents Dionysus supported by Eros preceeded by a muse with a lyre. *Courtesy of the Trustees of the British Museum.*

Greek folding (covered) mirrors were circular, polished metal discs often covered with silver and an ornamental frame, and could be carried or hung by a handle or hook at the hinge. The decoration developed in two ways. The exteriors of the covers were decorated in embossed or engraved reliefs, the edges often forming a framing border. The inner surfaces were sometimes ornamented with engravings, often with silver applied between the incised lines. Dionysian and erotic motifs predominate. Greek folding mirrors originated in workshops mostly outside of Attica. The most compact group points to Counth, but centers in Greece proper must also be considered. The portable folding mirrors only appear in Classical times, from about 450 B.C. to the fourth century B.C. From this period originate the most noble accomplishment in the elaboration of mirrors that has ever been achieved—at least in the west. (Hartlaub 185 and 186.)

7

8

7. Bronze mirror cover, Etruscan, third to second century B.C.

8. Bronze mirror back, Etruscan, fourth to third century B.C. The backs of Roman and Greek bronze and silver mirrors were beautifully decorated by symbolic elements. Here, a wreath of laurel leaves surrounds Ajax fastening his sword with the help of Teles (Thelius) while Acamena plays a lyre and crouching Selenus is drinking. *Courtesy of the Trustees of the British Museum.*

Aristotle speaks of "metal covered mirrors: when metal must be polished, to serve as a mirror, glass and crystal must have a metal coating, in order to reflect a picture". Pliny informs us that the Sidonians invented glass mirrors. Isidor of Seville, Pliny's echo, speaking of glass, says that there is no material more suited for mirrors. Pausaneas reports as a curiosity of a mirror in the temple of Diana in Arcadia at Naos, which was situated by the exit in the vicinity of the statue of the goddess. If one looked at oneself in it, one scarcely saw oneself, but the statue appeared very clearly. Seneca wrote that mirrors were produced in Rome to the height of a man and decorated with gold and silver, and even with jewels. (Kisa, p. 358)

9. Bronze mirror back, Bolsena, third to second century B.C. The engraved decoration on this mirror back represents the death of Troilus at the hands of Achilles and Ajax while Hector watches. *Courtesy of the Trustees of the British Museum.*

As early as 58B.C. mirrors were used on the walls of the Theatre of Scaurus in Italy. Also, mirrors were found in mosaic work. (Worshipful Company, p. 23)

9

Plate XI.

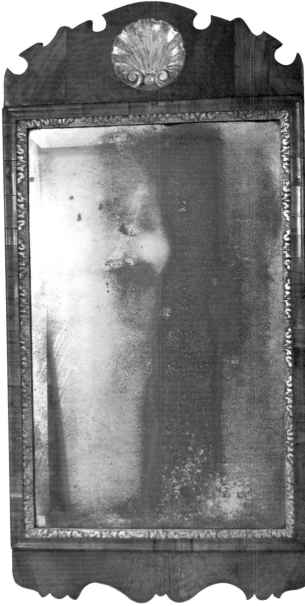

Plate XIII.

Plate XII.

Plate XI. Constitution or architectural mirror, walnut veneer, English, circa 1750, 53" high, 27" wide. see #224.

Plate XII. Detail of Plate XI showing the pediment carving and phoenix finial. Note how small the pieces of walnut veneer are. This is usually a sign of authenticity. Also note that a skillful cabinetmaker balanced the grain of the veneers.

Plate XIII. Queen Anne mirror, walnut veneer and gold leaf, English, circa 1730, see #103.

Plate XIV. Fretwork mirror, mahogany veneer and gilded wood, English, circa 1750, 43½" high, 25" wide, see #334. This gilding is very near its original, brilliant condition.

Plate XV. Fretwork mirror, mahogany veneer and gilded wood, English, circa 1750. Notice the intaglio carving which is an extra ornament and sign of refinement.

Plate XIV.

Plate XV.

10

Antiquity had knowledge of the use of mirrors as an optical aid to a high degree. There are reports of spying installations composed of mirror systems mounted by engineers on columns or towers.

The great installation on the lighthouse of Alexandria, the work of Alexander the Great, belongs to the history of antique technology. This was one of the wonders of the world. By using the mirror on this lighthouse, one could detect in advance the approach of hostile ships. (for further discussion, see H. Thiersch "Pharos", 1907.) (Hartlaub, p. 187)

A rich cycle of oriental and occidental legends has developed around the story of Pharos. (Hartlaub, p. 187)

10. Bone and glass, Egyptian, late fourth to fifth century A.D. 5¾" high, 4½" wide. Mirrored glass is known for reflective uses only since the late fourth century A.D. when the Romans had the technology to make small mirrors. Pliny the Younger mentions looking glasses, but none have survived from his period. This example is from the end of the Roman era in Egypt. *Courtesy of The Corning Museum of Glass.*

In Ireland, glassmaking probably goes back to the ninth or tenth centuries as some beautiful chalices of that date were found at Ardagh, Limerick. They are in the museum of the Irish Academy. (Worshipful Company, p. 36)

The amalgam of tin and quicksilver over glass was first used in Venice in making mirrors in 1317. (Worshipful Company, p. 30.) They had a monopoly in glass making all over the world until the end of the 17th century. By the middle of the 16th century the Venetian glass industry had a separate division of looking glass makers. The workmen lived and worked on the island of Murano. They were not allowed to leave. Punishment was death. The officials considered the glass making process secret and essential to their economy. Nevertheless, some

workers did get to England and France in the early 1600's.

Cast plate glass was made at Cherbourg, France, by a workman from Venice about 1300. The art was declared noble. (Worshipful Company, p. 30)

Glass gradually became a useful rather than ornamental material in Europe as the following points of information indicate. The first window of glass in England was installed in Wearmouth Church, County Durham, in 1440. In 1458, Aeneas Sylvius states, as proof of the wealth of the inhabitants of Vienna, that all houses had "Glass" windows. All the churches in France were supplied with window glass in the 16th century. Glass bottles were first made in England in 1537 and improved in 1635. Window glass was first made in England in 1557 at Crutched Friars under Cornelius de Lannoy (Worshipful Company, p. 31)

A mirror chamber of Torgan Castle is described by an "Antiquarian" of the 12th century as follows:

The mirror chamber, full of mirrors, formed in the most diverse manners, such that one could see above on the ceiling and on the walls around the table, in the room or in the bed, in the chamber, everything, which comes and goes, in the courtyard, in the street, in the country, and on the Elbe river.

That is, serving the "gallant" purpose as well as that of "Spying". (Hartlaub, p. 186.)

11. Convex mirror, Flemish, c. 1480. Mirrors were apparently unknown in Europe until the Renaissance when small examples were occasionally included as details in illustrations. This convex mirror was included in the Flemish manuscript "The Romant de la Rose". *Courtesy of The British Library.*

12. Oil painting, The Marriage of Giovanni Arnolfini and Giovanna Cenami, by Jan Van Eyck, 1434. The convex looking glass in this famous and amazingly detailed painting substantiates the presence of large convex mirrors in northern Europe in the fifteenth century. *Reproduced by courtesy of the Trustees, The National Gallery. London.*

11

Plate XVI.

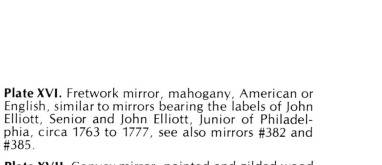

Plate XVI. Fretwork mirror, mahogany, American or English, similar to mirrors bearing the labels of John Elliott, Senior and John Elliott, Junior of Philadelphia, circa 1763 to 1777, see also mirrors #382 and #385.

Plate XVII. Convex mirror, painted and gilded wood and gesso, American, circa 1800.

Plate XVII.

13. and 14. Carved, painted, and gilded wood, Italian, c. 1500. The incredible carving in the frame emphasizes the importance at this time of the relatively small mirrored glass panel at the center. Letters, animals, exotic birds and intwined foliage are executed with precise detail. The back [14] is carved with cherubs and 2 adult figures. This mirror was owned by Lucrezia Borgia, the daughter of Pope Alexander VI, and one of the Renaissance's infamous, rich, beautiful, powerful and wicked women. *Courtesy of the Victoria and Albert Museum.*

16

15

15. Carved walnut, Italian, c. 1500. 31" high, glass 10½" high, 8¾" wide. *Courtesy of the Victoria and Albert Museum.*
16. Carved and gilded wood, Italian, c. 1560, 26" high, 24" wide. *Courtesy of the Victoria and Albert Museum.*
17. Carved walnut, Italian, sixteenth century. Each of these mirrors displays a different style of Renaissance design which surrounds and suports the valuable flat glass panels. The elements were re-utilized in the Classical period at the end of the eighteenth century by designers such as Robert Adam, and again in the Renaissance Revival period at the middle of the nineteenth century. *Courtesy of the Victoria and Albert Museum.*

18

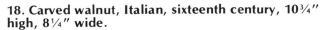

18. Carved walnut, Italian, sixteenth century, 10¾″ high, 8¼″ wide.

19. Carved walnut, partly gilded, European, six-teenth century. *Courtesy of the Metropolitan Museum of Art, bequest of George Blumenthal, 1944.*

20. Carved oak, European, sixteenth century, 19″ high, 12″ wide. The use of masks and human figures contorted to create the design of the frames were precursors of the Baroque style two hundred years later. They were modified but recognizable elements of the classical and Art Nouveau styles up to the twentieth century. However, in the Renaissance period the figures are carved with delicate precision yet lack individualized person-alities. They are idealized to the extreme of anonymity. *Courtesy of the Victoria and Albert Museum.*

19

20

20a. Steel and wood stand and frame, German or Northern Italian, late sixteenth century. Fine steel work was required for armour and sword making at this time. *Courtesy of Mrs. Madeleine Gimble.*

21. Convex mirror in wooden frame, Italian, sixteenth century. There is a striking similarity between this mirror and the one illustrated in the manuscript Romant de la Rose (see 11) made a century earlier. The lapse of years did not presume change in that era as it does today. Here additional applied rosettes were most certainly fixed to the frame originally. *Courtesy of the Victoria and Albert Museum.*

22. Carved wood, Italian probably Venetian, late sixteenth century. The design here utilizes figures and dolphins contorted to make the frame and stand around a small glass surface. The Baroque style is clearly emerging from the Renaissance. *Courtesy of the Victoria and Albert Museum.*

21

20a

22

23

23. Walnut inlaid with mother-of-pearl and painted gilt, Italian early seventeenth century. The growing commerce between Italy and the Middle Eastern region is reflected by the intertwined Persian design elements which appear on this mirror. *Courtesy of the Victoria and Albert Museum.*

24. Oil painting "The Toilet of Venus" or "The Rokeby Venus" by Diego Velazquez [1599-1660]. This type of mirror in rectangular frame was a very expensive luxury in Holland, England and most of Europe in the seventeenth century. The ribbons at the top were used to hang the mirror. The Renaissance style cherub is supportive of the composition, but no longer the main design element. Rather, the more Baroque sensuous female figure exhibits a definite personality and conscience through the use of the mirror. *Reproduced by courtesy of the Trustees, The National Gallery. London.*

When the Earl of Lester died in 1588 in England, his inventory of belongings included three mirrors, two of them "steele" and one "cristall".

No record of glassmaking exists in Scotland until 1620 when John Maria dell' Aqua from Venice was offered the post of Master of the Works in Scotland.

Sir R. Mansel, who had a patent for glassmaking in England, stated in November of 1624 "that he brought many expert Strangers from Forraigne parts beyond the Seas, to instruct the Natives in making all Sorts and Kings of right Christalline Morana glasses, and spectacle glasses, and looking-glass plates, and to wholly perfect the work." (Worshipful Company, p. 36) The English glass industry grew after The Worshipful Company of Glass Sellers of London was incorporated in 1664.

25

25. Walnut veneer, English, circa 1675, 38″ high, 21½.″ wide. The rectangular frame is convex molded with veneer in an oyster pattern made by cutting the wood at a 90° angle to the trunk, and arranging the pieces in alternate directions. Many rectangular mirrors of this period originally had an arched crest which may have become lost. Inspection of the mirror back to detect points where such an arch was joined is a customary procedure. *Courtesy of the Victoria and Albert Museum.*

26. Walnut and various inlaid veneers, English, late Seventeenth Century. The convex molded frame is decorated with scrolling leaves of contrasting wood inlay. This is a fairly typical example for this period when mirrors were still very costly and rare. *Courtesy of the Victoria and Albert Museum.*

Correspondence between John Greene and Michael Mesey of London and Allesio Morelli a glass maker in Venice between 1667 and 1672 has been preserved and reprinted by The Worshipful Company of Glass Sellers in London. This exchange verifies that large quantities of both table (drinking) glass and ornamental (mirror) glass were transported at that time from Venice to London. The letters also reflect the development of the English glass business at the time and the government's attitude toward tariffs meant to protect the growing English industry.

In February, 1670-71, Messers Greene and Mesey wrote that since a previous order was unsatisfactorily received, they hoped "for satisfaction

26

27

(with the next order) or else truly you will discourage me for the future to send to you (or any other person at Venice) for any more Looking Glasses since we have so good made here in England, and so many also of Venice Looking Glasses in London to be sold so cheap, for I assure you I have bought some cheaper here than my last I had from you".

In November, 1672, John Greene wrote to Mr. Morelli "Pray let all these (Looking Glasses) be packed in the bottoms of the chests of drinking glasses that I may thereby save the custom of them here, and to that end I must further desire you to send me two factorys (packing slips?) and in one of your factories not to mention any looking glasses nor the full number of the drinking glasses by 8 or 10 dozen in a chest for our officeres here are so strict that they will see our factories."

27. Walnut, ebony and contrasting veneers, English, circa 1675, 50" high, 39" wide. A later stylistic decorative device than continuous scrolled leaves is the grouping of the decoration in reserves. This is found on Dutch as well as English mirrors of the time. *Courtesy of the Victoria and Albert Museum.*
28. Oil painting "A Merry Company at Table" by Hendrick Pot Dutch, circa 1585. The rectangular molded frame mirror hangs in the home of a very wealthy person of the period. The clothes shown distinctly represent the upper class. *Reproduced by permission of the Trustees of the Wallace Collection.*
29. Oil painting "A Woman Peeling Apples," by Pieter de Hooch, Dutch, circa 1670. The mirrors shown in these paintings are a typical size and style for the period. *Reproduced by permission of the Trustees of the Wallace Collection.*

28

30

31

32

BAROQUE

In 1677 the Duke of Buckingham brought Venetian workmen to Lambeth 'where they made huge vases of metal [glass], as clear, thick, and ponderous as crystal, also looking-glasses far finer and larger than any that come from Venice.' (Evelyn)

30. Oil painting, by Hendrick Pot, Dutch, late sixteenth century. *Reproduced by permission of the Trustees of the Wallace Collection.*

31. Ebony inlaid with ivory and contrasting wood, English, circa 1675. This mirror hangs as it always has at Ham House. The pattern is of jasmin flowers and European style scrolled leaves. There is an inlaid face in the center of the crest below a ducal cornet. The cornet dates the mirror after the creation of the Lauderdale Dukedom in 1672. It is believed that this is the mirror listed on the Ham House inventory of 1679, making its acquisition probable over the seven year period. The scalloping along the top crest is a detail which continued into the Queen Anne and Chippendale periods. *Courtesy of the Victoria and Albert Museum.*

32. Walnut and contrasting veneers, English, late seventeenth century, 41" high, 26 5/8" wide. The simple cross banded outline sets off the inlaid decoration on this mirror. The tight floral design is similar to the Dutch marquetry of this period.

33. Japanned wood and quill work, English, circa 1690. Oriental lacquered furniture was imported to England by the East India Company after 1660 and proved so popular that in 1688 John Stalker and George Parker wrote a *Treatise of Japanning and Varnishing* giving instructions to their countrymen. This mirror is probably a result of attemps by Englishmen to make their own Japanned articles. The Japanned frame supports the mirror as well as panels decorated with quill work which is gilded paper rolled, glued and arrange in intricate patterns. *Courtesy of the Victoria and Albert Museum.*

34

35

36

37

38

34. Oak veneered with Oriental lacquered wood, English, circa 1675, 60" high, 39" wide. One type of Oriental lacquer ware is "Bantam work" which has incised and colored decoration on the lacquered boards. In England, lacquered screens and planks were cut up and used to decorate furniture such as this mirror. Here the original pattern is interrupted and turned to create a lively, if confusing, decoration. *Courtesy of the Victoria and Albert Museum.*

35. Walnut veneer on pine, probably English, circa 1700, 46" high, 28½" wide. This otherwise standard mirror of the period is embellished with three raised ovals or "bosses'" which were sometimes applied to furniture in England and New England in the late seventeenth century. A picture of this particular mirror was used on the bookplate of Henry Francis du Pont, the founder of the Winterthur Museum. *Courtesy of the Henry Francis du Pont Winterthur Museum.*

36. Walnut veneer, English, late seventeenth century. The decorative effect of inlay is achieved here by cut out spaces creating the contrasting light and dark areas. This is a highly unusual example of William and Mary style mirror. The royal insignia of crown, lion and unicorn are brilliantly incorporated into the design. *Courtesy of Price Glover, Incorporated.*

37. Painted wood, American, New England, circa 1700, 46¾" high, 20½" wide. The pierced cresting of the previous example is interpreted differently on this mirror which is also painted with a marbled pattern. During the William and Mary and Queen Anne periods, painted marble decoration was used in New England on mirrors. *Courtesy of the Henry Francis du Pont Winterthur Museum.*

38. Walnut veneer, English, circa 1700. The interesting pattern of the veneer is highlighted on the crest which is also ornamented by a scalloped outline. *Courtesy of Herbert Schiffer Antiques.*

39. Walnut veneer, English, circa 1690, 30" high, 27" wide. This mirror marks a transition in the evolving design styles from William and Mary (the square proportion and molded frame) to Queen Anne (the use of serpentine outline in the crest). *Courtesy of David Stockwell, Incorporated.*

40. Lacquered wood and stumpwork embroidery, English, late 17th century, 28" high, 24" wide. This mirror is in a beautiful state of preservation. The lacquered wood frame is painted with a gold design copying Japanned wood from the Orient while the oval glass panel is an unusual shape, as most mirrors of this date are rectangular.

The stumpwork embroidery which surrounds the glass was the work of young girls and professional embroiderers to show off their needlework skills. It has retained much of its original coloring. Typical motifs are present—the queen under a canopy, duck, lion, unicorn, overscale flowers and bugs. The pattern is drawn to the background and padding and colored threads are woven to create a raised design. Considering the age and delicacy of this material a large number of stumpwork relics remain today, perhaps because they have always been cherished and protected. *Courtesy of the Victoria and Albert Museum.*

41. Straw work, English, late seventeenth century. This mirror probably had an outside wooden frame similar to the preceeding example. A variation of stumpwork decoration is straw work as presented here with the same general group of motifs popular in the period. The straw pieces were cut to the required size and bleached or dyed before being glued to the background. The result is a flatter design than stumpwork. The process apparently originated in Italy before it spread to Spain, France, and finally England. *Courtesy of the Victoria and Albert Museum.*

42. Tortoise shell and stumpwork embroidery, English, late seventeenth century, 24½" high, 21 3/8" wide. The stump work background may have been intended for a frame conforming to its outline. Instead, this rectangular mirror is framed with tortoise shell in a shape sympathetic to the outline, but squared. The arching crest shows a gradual change in taste toward the Queen Anne style. *Courtesy of Mrs. Herbert F. Schiffer.*

41

43

43. Wood and embroidered needle-work, English, late seventeenth century. The straight molded frame acts only as a support for the needlework panels which surround this mirror. No longer is the wood itself decorated.

44. Beadwork on wood backing, English, dated 1662. Beadwork is another variation of embroidery or stumpwork used in the late seventeenth century to ornament small articles and mirror frame panels. This example retains its original bright colors and probably was itself surrounded by a frame. *Courtesy of the Victoria and Albert Museum.*

45. Traveling mirror covered in silk embroidery, English, late seventeenth century, 11" high, 6 3/8" wide. The embroidered panels completely enclose this mirror, including two swinging panels to cover the glass. The design reflects the popular taste for intertwined flowers and leaves. As this was designed for use while traveling, it was probably further protected by an outside wooden box. *Courtesy of Mrs. Herbert F. Schiffer.*

46 and 47. Coffer of silk stumpwork over wood backing, English, circa 1680, 13" high, 14¾" deep, 9" deep. Coffers of this type were used as jewelery and make-up boxes. Many of them have ingenious secret drawers. The interior mirror is removable in this example. *Courtesy of Mrs. Herbert F. Schiffer.*

45

47

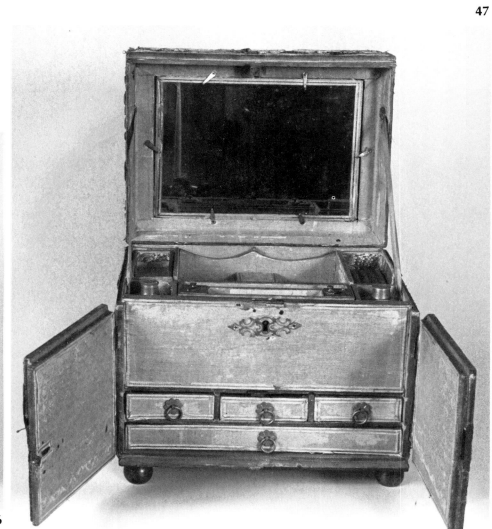

46

48

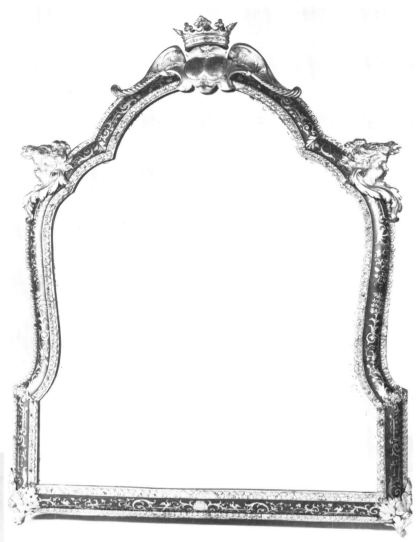

50

51

48. Verre Eglomise, The Penitent Magdalen, possibly French or Italian, circa 1670. The two mirrors shown in this picture are symbolic of Vanity, the hanging one on the wall tempting the Magdalen to vanity, and the reclining one on the floor displaying her renunciation of vanity. The picture itself is a fine example of Verre Eglomise technique using etched gold and silver foil behind the clear glass reverse painted picture. The frame in this example also displays fine Continental Baroque styling. *Courtesy of the Victoria and Albert Museum.*

49. Carved walnut, French, late seventeenth or early eighteenth century. This dressing mirror was made as part of a toilet set with such detailed carving as to resemble repousse silver, the more common and expensive frame material for dressing mirrors of this period. The glass itself is beveled to conform with the outline of the glass.

50 and 51. Tortoise shell and engraved brass over wood, French, late seventeenth century. Andre Charles Boulle devised finely engraved brass inlay in toroise shell as a decorative ornament while working under Louis XIV in France. This technique, known as Boulle work, was one of the most luxureous products of this age and was copied, but never as beautifully, in subsequent revival periods. The frame and back of this dressing mirror display the technique magnificently. *Reproduced by permission of the Trustees of the Wallace Collection.*

Inuenté par D. Marot.   auec Preuillege des Etats Genraux des Prouinces Vnie &   a Ollande d. W. S.

52. Four designs for mirrors by Daniel Marot, French, late seventeenth century. Daniel Marot was a Huguenot and designer for Louis XIV. He left France for Holland in 1684 when revocation of the Edict of Nantes meant persecution for the Huguenots. If he himself did not get to England soon thereafter, his designs most certainly were used to inspire English mirrors made very closely resembling these designs. In the following examples, silver mirrors from the collections of the English royal families show the inspiration of Daniel Marot's designs. *Courtesy of the Victoria and Albert Museum.*

53. Gilded wood, Dutch, late seventeenth century. This Dutch mirror very closely reflects Daniel Marot's designs. The carving is beautifully detailed. Daniel Marot went to Holland from France. His design book went to England when Charles II was restored. *Courtesy of M. Turpin, Inc.*

54. Oil painting, Dutch, signed N. Maes, 1656. The mirror and doorway arch both are designed in the Baroque style and display the fashionable taste in domestic interiors of the mid-seventeenth century. *Courtesy of the Victoria and Albert Museum.*

53

54

55

56

57

**55 and 56. Silver, English, circa 1684, mirror 13"
high, 17" wide.** Very elaborate silver dressing sets
were first made in France. So much silver was being
used for dressing sets there that between 1672 and
1678, Louis XVI prohibited the sale of silver mirrors.
Silver was needed, instead, for coinage to pay for
his wars. At this time the use of gold leaf on
furniture in the King's court became popular
because little gold was actually needed to create an
enormously expensive appearance. After the revocation of the Edict of Nantes in 1684, French
Huguenot silversmiths brought silver chasing techniques to Germany and England. The silver sets
were all enormously expensive and therefore made
only for the richest customers. This is the Calaverly
set, made in the manner of Daniel Marot's designs
shown previously. The crest detail in picture 56
illustrates the finely chased design and high relief
for which the makers became so well known.
Courtesy of the Victoria and Albert Museum.

57. Silver, English, circa 1675 to 1680, King Charles
II presented this dressing set to the Strickland
family, formerly of Sizergh Castle, Westmorland.
The pieces are each engraved in the "Chinese taste"
which complemented Oriental lacquer ware furniture which was very popular at this time. *Courtesy
of the Victoria and Albert Museum.*

**58. Silver over oak, English, by Andrew Moore of
Bridewell, circa 1695, 90" high, 47" wide.** This is the
most magnificent mirror the author has seen, no
picture does it justice. It was presented by the city
of London to King William III along with a matching
pier table and two candlesticks. The detail of the
silver chasing is crisp and each element clearly
defined in opulent profusion. The Royal Crest is
flanked by lion and unicorn with surmounted
crown. *By gracious permission of Her Majesty the
Queen.*

**59. Oil painting. Toilet scene from the series
Mariage a-la-mode, by William Hogarth (1697-
1764), painted circa 1738.** A wonderful late
seventeenth century silver dressing set is shown in
use in this famous painting. *Reproduced by
courtesy of the Trustees, The National Gallery,
London.*

58

60

60. Silver over oak, English, circa 1670, 83" high, 50" wide. The City of London presented this mirror en-suite with a pier table and two candlesticks to King Charles II. A further embellishment in the form of a crown above the crest probably originally existed, such as those in the drawings by Daniel Marot. The cartouch bears the King's initials "C R" (Charles Rex) and superb repousse work appears in the cherubs, bunches of grapes, foliage and ribbons on the frame.

John Evelyn, the diarist mentions seeing gold mirrors at Hampton Court and silver mirrors in Genoa. We can assume today that each of the wealthy European courts had silver mirrors, as they are also mentioned by visitors to the French court of Louis XIV.

After the French forbade the use of real silver and gold for furniture decoration (as it drew too deeply from the supply needed to finance their wars), gold and silver gilding became customary at court. *Reproduced by gracious permission of her majesty the Queen.*

62

61. Silver gilt, English, circa 1670, 69½″ high, 47″ wide. Very few original seventeenth century silver gilt mirrors have survived. On this example, the heavy square frame and thick carving are characteristic of the period's general quality. This mirror retains its original beveled glass panel. The coat of arms at the top represents the Perry family of Perry Hall, Staffordshire, impaling Gough of Old Fallings and was granted in 1664. *Courtesy of the Victoria and Albert Museum.*

62. Walnut and silvered metal, English, circa 1710-1715, 76″ high, 52″ wide. The period of silver mounted furniture was short but glamorous. This example of a suite of matching furniture was very popular with the wealthy English as it was one step less costly than ones entirely covered with silver, but very few examples have survived. This group is displayed at Ham House, Richmond, England. *Courtesy of the Metropolitan Museum of Art.*

63. Carved and partially gilded pine, English, circa 1680, 43¼″ high, 31″ wide. The carving is of strong, design and well executed. In the crest, the cherubs hold a shield with the arms of Hildyard, and in the base a cypher is cut-out with delicate scrolls. *Courtesy of the Victoria and Albert Museum.*

63

64. Silver and gold leaf, English, circa 1680, 66" high, 53" wide. The cherubs on this mirror are silver gilt while the remainder of the carving is gold leaf. It represents a transition between gilt-silver and entirely gold leaf mirrors, and therefore is a most interesting and highly unusual mirror. The tendency was toward fewer cherubs and more gold leaf.

65. Carved wood, Italian, late seventeenth century. This mirror possesses many of the elements which are the prototypes for French and later English Rococo designs. Scrolling leaves, elaborate crests, and finely carved and pierced symmetrical frames became fashionable and took on their own national characteristics. The designs of Daniel Marot were obviously influenced by this type of predecessor. *Courtesy of the Metropolitan Museum of Art, Bequest of Theodore M. Davis, 1915.*

66. Carved limewood by Grinling Gibbons, English, circa 1680. 15½" high, 12½" wide. Grinling Gibbons carved this oval mirror frame early in his career, perhaps soon after he was "discovered" about 1671 carving brilliantly by John Evelyn the diarist who brought him to the attention of the King. This boy of English parents was born in Holland, and apparently was influenced by naturalistic but carefully planned Dutch paintings of flowers. This mirror was probably made for the decorations at Cassionbury Park, Hertfordshire, a private house which no longer exists but some of its interior is preserved at the Metropolitan Museum of Art, New York. The punched background of the mirror frame sets off the softness of the carved flowers, pea pods and wheat. *Courtesy of the Victoria and Albert Museum.*

65

66

44

67. Carved walnut, English, attributed to Ginling Gibbons, late seventeenth century. This frame is a later and even more ambitious example of Grinling Gibbons' carving than the preceeding mirror. The changes in texture, plane and combination of details is truly impressive. The hard sea shells and crab at the base contrast with the wilted leaves and blossoms at the crest. His use of pea pod details is almost a signature. Lime wood, pear wood and walnut were the principal materials for carved frames at this period and up until the mid-eighteenth century when American white pine was sometimes brought to England.

It is difficult today to precisely identify frames for mirrors and those for paintings unless the original old glass panels remain. *Courtesy of the Victoria and Albert Museum.*

Glass framed

68

69

68. Gold leaf, English, circa 1700. An important group of early eighteenth century mirrors have glass paneled frames and simple arched crests. Some of these have eglomise or blue glass panels in the frames.
69. Gold leaf, English, early eighteenth century. The cut decorations in the frame panels and arched crest became typical of the early eighteenth century period in England. *Courtesy of M. Turpin Antiques.*

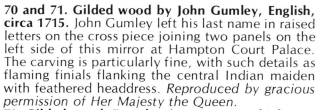

71

70 and 71. Gilded wood by John Gumley, English, circa 1715. John Gumley left his last name in raised letters on the cross piece joining two panels on the left side of this mirror at Hampton Court Palace. The carving is particularly fine, with such details as flaming finials flanking the central Indian maiden with feathered headdress. *Reproduced by gracious permission of Her Majesty the Queen.*

72. Gilded wood, French, circa 1680, 86" high, 51" wide. The early date of this mirror illustrates the superior manufacturing skills of the French to make large glass plates earlier than the English. The style of mirror may be seen as a prototype of the English of the early eighteenth century, as so much of the French taste inspired later English styles. This French mirror possesses a more exuberant decoration in scrolled leaves than comparable English mirrors.

72

73

74

73. Gilt and Eglomise, English, circa 1700. *Courtesy of the Victoria and Albert Museum.*

74. Eglomise, English, circa 1700, 84" high, 31¾" wide. These mirrors closely relate to Italian Renaissance antecedents with tightly designed trellis decoration in the eglomise frame panels. The carved and pierced crest of #73 is translated entirely in eglomise on #74. *Courtesy of the Victoria and Albert Museum.*

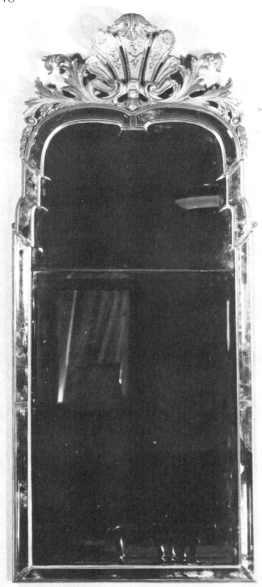

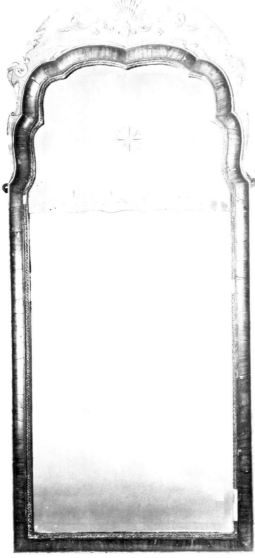

75

77

76

78

75. Gilded wood, English, circa 1695. This mirror style bridges the transition between the mirrored frames and gilded frames as it incorporates both elements. It is rare to find a mirror of this style today.

76. Walnut veneer and gold leaf, English, circa 1720. Another step in the transition toward all-gilt mirrors is this style with walnut veneer in place of the mirrored frame, still with cut decoration in the top glass, beveled lower glass, and a carved and gilded crest. *Courtesy of the Victoria and Albert Museum.*

77. Walnut veneer and gold leaf, Danish, eighteenth century. Some of the fine mirrors which were made in the Scandinavian countries are later in date than their English counterparts as there was a cultural lag in the developments of styles. This same lag is found in American styles. Although similar to the English mirrors in this progression of styles, this mirror was made about 1770. The carving is heavier, or bolder, than the English mirrors of this date, and the winged figures are closer to the German folk art tradition than to French or English prototypes. *Courtesy of M. Turpin Antiques.*

78. Glass and brass, Dutch, early eighteenth century, 36" high. The cut decoration on this beautifully shaped, large glass sconce is interesting and desirable. It was probably made with many other of the same design for the same room. Sconces of this type were also made elsewhere, but not as large. *Courtesy of David Stockwell, Inc.*

79

Molded

82

81

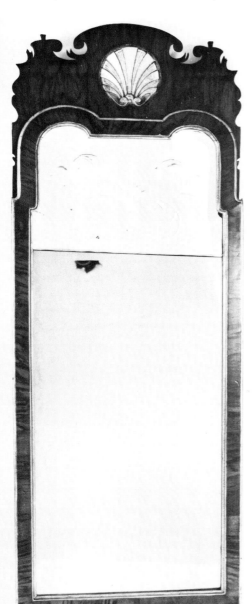

80

79. Walnut veneer, English, early eighteenth century. It is unusual to find a complete mirror of this shape, which corresponds to the upper plate of longer mirrors of the period. It may have been made for use over a small mantle or high chest of drawers to reflect light. The interesting cut pattern of the original glass is of fine quality and the convex frame is outlined by the beaded edge. *Courtesy of Herbert Schiffer Antiques.*

80. Walnut veneer and gold leaf, English, circa 1720. The veneered frame of this mirror is flat and the outline and fillet are gilded. The entire crest is removable, a feature which probably existed on many mirrors originally but which have become lost. *Courtesy of Herbert Schiffer Antiques.*

81. Japanned wood and gilded decoration, possibly American, circa 1730, 41½" high, 20" wide. The Schuyler family of the Hudson River Valley region of New York State owned this mirror which may have been made locally. In comparison to the quality of Japanning work being done in England, the decoration here is mediocre at best. Japanned work of this type is known to have been made around Boston, Massachusetts where one craftsman, Thomas Johnson worked as a "Japanner at the Golden Lyon in Ann Street Boston" between 1732 and 1767. *Courtesy of David Stockwell, Inc.*

82. Japanned maple and pine with gilded decoration, American, circa 1720-1745, 47½" high, 19 3/8" wide. The secondary woods verify this mirror to be of American origin probably Boston, Massachusetts. The decoration is relatively primitive when compared with Japanned work from England or the Orient, however the glass is beautifully cut and there is no doubt of its European origin. *Courtesy, The Henry Francis du Pont Winterthur Museum.*

84

83

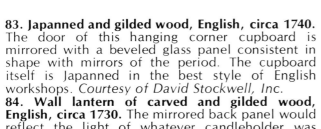

83. Japanned and gilded wood, English, circa 1740. The door of this hanging corner cupboard is mirrored with a beveled glass panel consistent in shape with mirrors of the period. The cupboard itself is Japanned in the best style of English workshops. *Courtesy of David Stockwell, Inc.*

84. Wall lantern of carved and gilded wood, English, circa 1730. The mirrored back panel would reflect the light of whatever candleholder was placed behind the curved glass sides. This is a fine example of a rare hanging wall lantern.

85. Wall lantern of mahogany, English, circa 1720-1730, 25″ high, 12½″ wide, 9″ deep. This form of lantern is mentioned in an advertisement for the September 7, 1719 *Boston News Letter* as being imported to Boston. *Courtesy of Israel Sack, Inc. New York City.*

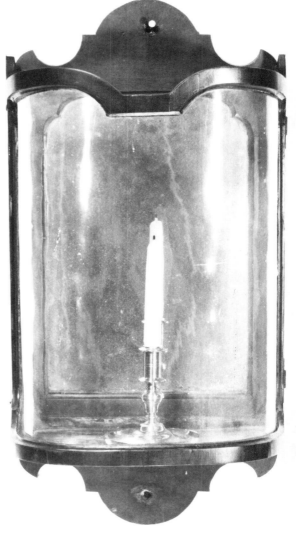

85

86

87

88

89

86. Walnut veneer, English, early eighteenth century, 32½″ high, 20½″ wide.
87. Walnut veneer, English, early eighteenth century. *Courtesy of Herbert Schiffer Antiques.*
88. Mahogany veneer, English, early eighteenth century.
89. Mahogany veneer, English, circa 1720 to 1760.
The shape of mirrors evolved through the early eighteenth century to a standard long glass with curved upper edge and simple molded frame. The crest became taller, gradually, and with even-more elaborate cut-out shaping. These characteristics collectively define the Queen Anne style in mirrors which will be shown in many variations in the following examples.

While the great majority of this style are of English origin, American-made mirrors of this Queen Anne style have been identified by wood analysis. Both mahogany and walnut primary woods are found. The firm of John Elliott imported from England and sold this style in Philadelphia in the 1750's. *Courtesy of Herbert Schiffer Antiques.*

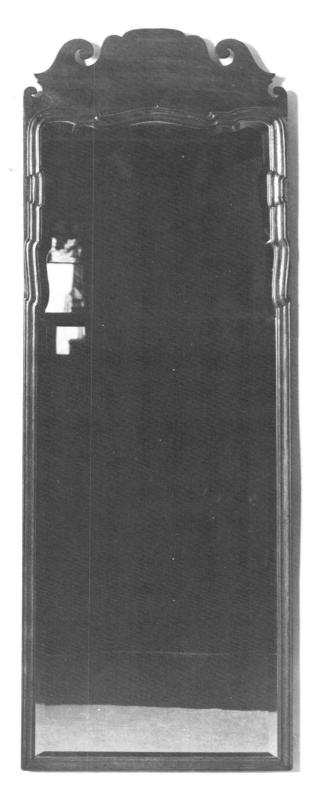

91

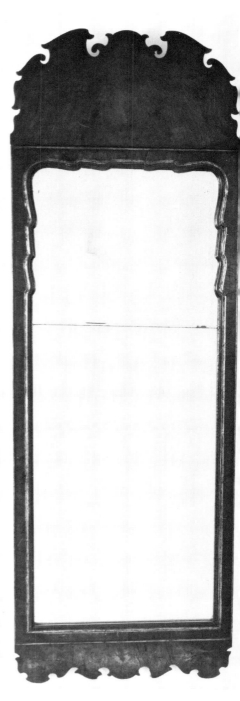

92

90

90. Mahogany veneer, English or American, circa 1740. 67" high, 23" wide. This extremely tall mirror was owned in the eighteenth century by Jeremiah Cresson of Philadelphia whose initials are on the back of the mirror. *Courtesy of David Stockwell, Inc.*

91. Mahogany, English, circa 1730-1740. *Courtesy of Herbert Schiffer Antiques.*

92. Mahogany, English, circa 1740. *Courtesy of Herbert Schiffer Antiques.*

93

94

95

93. Sconce, walnut veneer and gold leaf, probably English, circa 1740, 28½" high. The shell in the crest is intaglio carved and gilded. The back plate of the sconce arm is unusual in its oval shape as one would expect a Baroque shape at this period. *Courtesy of The Detroit Institute of Arts, Gift of Mr. and Mrs. George F. Green.*

94. Walnut veneer and gilded, carved wood, American, circa 1740-1760, 60½" high, 21 7/8" wide. This mirror was made in Philadelphia. The applied shell and leaves make an unusual crest decoration and are similar to carving found on Philadelphia-made mantle pieces and case furniture. *Courtesy of Israel Sack, Inc. New York City.*

95. Mahogany veneer and applied gilded wood and brass sconce arms, Danish or other Scandinavian, circa 1740. This mirror is a superb example of the Scandinavian mirrors which were imported frequently to America in the eighteenth century. *Courtesy of David Stockwell, Inc.*

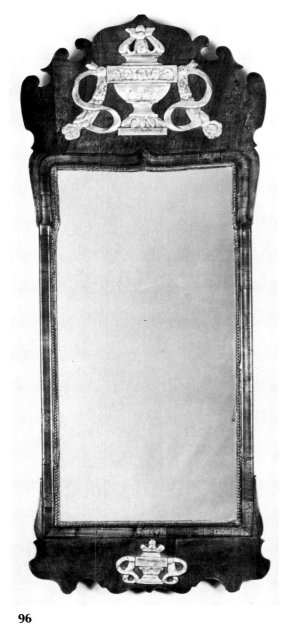

96

97

98

96. **Walnut veneer on pine with applied gilded wood. Danish or other northern European, circa 1770, 41" high, 16¾" wide.** *Courtesy of James Billings.*

97. **Rosewood and applied gilded wood, Portuguese, mid-eighteenth century, 57" high, 26" wide.** *Courtesy of Dillingham and Company.*

98. **Walnut veneer and gold leaf, English, circa 1730, 41" high, 16¾" wide.** These three mirrors have strikingly similar overall design, yet are from three distinctly different areas. Each has crest ornaments consistent with its region's own artistic traditions while the Common French through Italian design source is evident. The rounded upper edge of the glass plates, frame moldings, and overall long shapes are shared details.

99

100

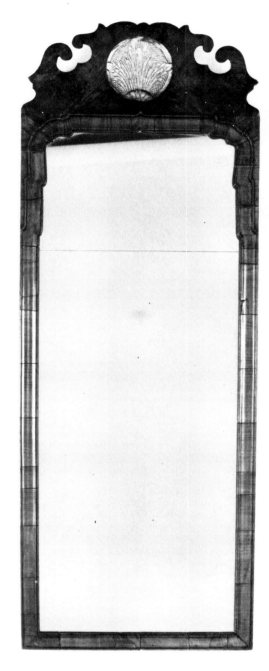

101

99. Walnut veneer and gold leaf, English, circa 1730, 48" high, 18½" wide.
100. Mahogany veneer, English, circa 1730.
101. Walnut veneer, English or American, circa 1730. The overall tall shape and graceful moldings and crest outlines on each of these mirrors are good examples of "Queen Anne" style. Each of the gilded shells is intaglio carved, but of different designs. *All three courtesy of Herbert Schiffer Antiques.*

102   **103**   **104**

**102. Mahogany veneer and gold leaf, English, circa
1730.** *Courtesy of Herbert Schiffer Antiques.*
**103. Walnut veneer and gold leaf, English, circa
1730.** *Courtesy of Herbert Schiffer Antiques.*
**104. Walnut veneer and gold leaf, Danish or
Northern European, circa 1730 to 1780.** *Courtesy of
Edward Steckler.*

105

105. Three painted frames, probably American, mid-eighteenth century. An infinite variety of designs and carving are found on the crest decorations of these Queen Anne mirrors which occur with walnut or mahogany veneer.

In the eighteenth century, there were made many more small mirrors than large. Small ones sometimes were made from the broken pieces of large mirrors because mirrored glass was very expensive. Also, small glass panels could be blown while large panels had to be cast. Museums have not collected many small mirrors, yet as late as thirty years ago, old homes in the Pennsylvania and New England countryside commonly had old ones in use. In Philadelphia and the larger cities, important rooms of the great houses had larger mirrors with more elaborate crest decorations.

The two outside mirrors in picture 105 have black painted frames with gold Chinoiserie decorations. This decoration was found in the 1720 to 1750 period in New England and was not refined, but an approximation of Oriental design. The central mirror is a rarity because of the boss in the crest. *Courtesy of Herbert Schiffer Antiques.*

106. Mahogany veneer, Danish, circa 1740. *Courtesy of Herbert Schiffer Antiques.*

107. Left, painted wood. Center, mahogany and eglomise. Right, painted wood and eglomise. All Northern European, mid-eighteenth century. Each of these small mirrors is made in a different stylistic tradition, yet they are contemporary examples of mid-eighteenth century design. Picture 106 shows a type now given a Danish origin, but once thought to be American. Picture 107, left, represents the Venetian tradition of ornate Baroque carving. The center one is an early "Courting" mirror described on the following page. The right is a predecessor of "Courting" mirrors in that the carved crest (in Baroque tradition) surrounds a reverse painted panel (eglomise). *Courtesy of the Metropolitan Museum of Art, gift of Russell Sage.*

106

108. Eglomise in original pine box, European, eighteenth century, 20½" high, 17½" wide. *Courtesy of David Stockwell, Inc.*

109. Eglomise and wood, Northern European, eighteenth century, 15½" high, 10½" wide. *Courtesy of Munson-Williams-Proctor Institute.*

110. Mahogany veneer and eglomise, European, eighteenth century. *Courtesy of Herbert Schiffer Antiques.*

111. Walnut veneer, pressed brass gilded fillet, and eglomise, European, eighteenth century. *Courtesy of Mrs. Herbert F. Schiffer.*

These "Courting" mirrors were made apparently throughout the eighteenth century. Some examples very closely relate to seventeenth century veneered styles with convex molded frames (such as picture 110) while others have a great deal of glass in the frames (picture 109) like the large early pier mirrors.

As a group, many have survived perhaps due to the original wooden carrying boxes in which they were kept. It is unusual today to find the boxes intact. The term "Courting" is as uncertain in origin as the mirrors are themselves. A surprisingly large number of courting mirrors have descended in old American families, which led previous historians to guess that they may be of American origin, but no conclusive evidence has thus far been found. This author believes them to be from Northern Europe.

108

109

110

111

Dressing

112. Oil on canvas painting "The Broken Mirror" by Jean-Baptiste Greuze [1725-1805], French, mid-eighteenth century. The broken glass in this small dressing mirror was certainly valuable enough to cry about. *The Wallace Collection.*

60

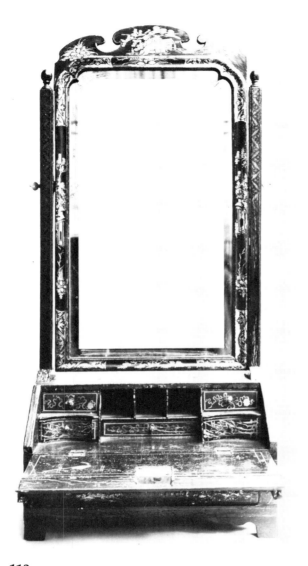

113

114

115

117

116

118

113. Painted pine and gold leaf, English, early eighteenth century, 33" high, 15¼" wide. Fascination with the Orient extended from the late seventeenth century trading missions right up through the nineteenth century. The painted decoration on this dressing mirror frame and case are English fantasies in the Chinese manner, a very popular decoration. The case on this mirror resembles a slant front desk and interior. *Courtesy of the Victoria and Albert Museum.*

114. Mahogany and brass inlay, English, circa 1740. The carving and extremely rare brass inlay are botho of the highest quality making this dressing mirror an exceptional piece. *Courtesy of the Victoria and Albert Museum.*

115. Oil on canvas painting "La Modiste" by Francoise Boucker [1703-1770], French, mid-eighteenth century. The dressing mirror in this painting is typical for the period in elegant homes. Charming French paintings of lovely women in their dressing chambers were common, and the intimate glimpses of every day life they provide later students is vastly interesting. William Hogarth in England was also a keen observer who included interior views with mirrors. Under his hand, the painting turns to satire and caricature.

116. Mahogany and mahogany veneer, English, circa 1730, 17¼" high, 9½" wide. The trestle feet on this charming, small mirror are quite unusual. The two side finials and wing nuts are hand wrought brass. *Courtesy of Herbert Schiffer Antiques.*

117 and 118. Walnut veneer, English, circa 1720, 11½" high, 6¾" wide. This mirror seems to be in the tradition of courting mirrors with its angular convex molded frame and tall crest. The easel-like stand is quite unusual with old leather thongs an interesting hinge idea. The overall form reminds one of the early French and Italian mirrors (see picture 32). *Courtesy of Mrs. Herbert F. Schiffer.*

119

119. Walnut veneer on oak, English, early eighteenth century, 32" high, 17" wide. This is an exquisite example of what became a common style of dressing mirror. The choice of highly contrasting wood makes an attractive surface and the applied heart-shaped boss is an endearing detail. *Courtesy of the Victoria and Albert Museum.*

120. Walnut veneer and gold leaf, English, circa 1720. Many fine quality dressing mirrors from this period have survived. This one has particularly nicely chosen wood, a central drop in the skirt which is a design vestige of case furniture of the seventeenth century, and pleasing crest shape and gadrooned edge band. *Courtesy of the Victoria and Albert Museum.*

121. Mahogany with oak secondary wood, English, sold by and bearing the paper label of John Elliott of Philadelphia, circa 1760. Stylistically this dressing mirror is akin to the other Queen Anne examples in this section of the first quarter of the eighteenth century, yet the style persisted especially in America well into the third quarter of the century. The paper label of John Elliott fixes the original sale to the 1760's. Call it cultural lag or a continuity of style, but America held on to the "Queen Anne" long after the Rococo "Chippendale" style was popular in Europe. *Courtesy of David Stockwell, Inc.*

122. All glass, English, circa 1695. Very few overmantle mirrors of this style have survived until today, and this mirror is a particularly fine example with finely beveled and shaped borders. Glass framed mirrors persisted in England until about 1720, and in Ireland and Italy until later in the eighteenth century.

120

121

122

Overmantle

123. Blue glass borders, English, circa 1700, 50" high, 75" wide. The blue glass borders emphasize the architectural elements of this overmantle mirror. The nails joining the small border panels to the frame are incorporated into the decoration and covered by the cut, clear, mirrored, floral glass pieces. This mirror hangs in Hampton Court Palace where the room molding was arched to accommodate the mirror height. *Reproduced by gracious permission of Her Majesty the Queen.*

124

124. Tortoise shell over wood, English, circa 1720-30, 63" long, 19½" high. The original Vauxhall cast plate glass panels remain within this tortoise shell covered frame. In this and like mirrors the design is inconsequential, allowing for the exotic tortoise shell to be utilized and made the important feature. *Courtesy Tryon Palace Restoration, New Bern, North Carolina.*

125. Eglomise and gilded wood, English, circa 1700. As with the previous mirror, the design of this frame is secondary to its decoration. The eglomise is colored gold, red, blue and green in an Italian Renaissance design.

125

126

127

126. Gilded wood, English, early eighteenth century. The carving, punching, and side shaping of this overmantle mirror are good examples of the gilded style of which many vertical examples will be seen in the next section of this book. The undecorated reserves in the lower corners are where sconce arms are meant to be attached.

127. Gilded wood and gilded pewter, English, circa 1735, 24½" high, 62" wide. This is a fine gilded overmantle mirror made even more interesting by the gilded pewter inner fillet. Such pewter details were also placed on the spandrels of clocks during this era.

128. Walnut veneer on pine, English, circa 1725. Here is the overmantle version of the vertical fret-work mirrors of which so many examples will be seen in a later chapter of this book. The three panels of glass are each beveled on four sides and the candle arms are original designs which cast a great deal more light from the mirror into the dark chamber. *Courtesy of the Metropolitan Museum of Art, Gift of Luke Vincent Lockwood.*

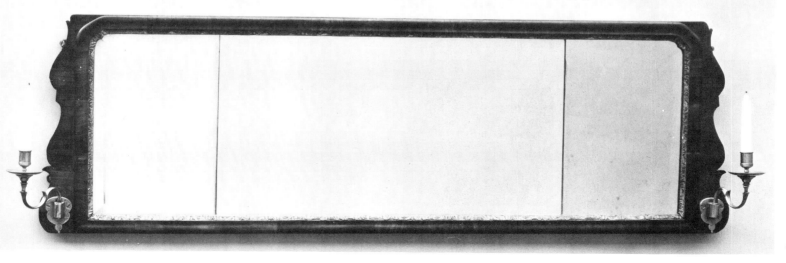

128

129

130

131

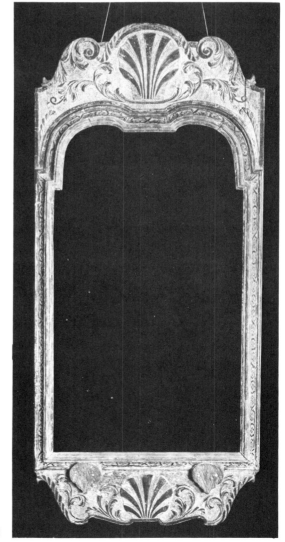

132

133

134

Gilded

129. Gilded mirror, Venetian, probably seventeenth century, on mahogany and pine base, American, circa 1760 to 1780, 28¾" high, 18¾" deep. The much older European gilded mirror was later mounted on a fine piece of American cabinet work, probably made in Massachusetts. This combination is, of course, probably unique, and it emphasizes the care which mirrors were given, particularly an old and cherished one. *Courtesy, The Henry Francis du Pont Winterthur Museum.*

129a. Gilded wood, English, circa 1710, 32" high, 19" wide. The simple gilded frame sets off the architectural design elements of this style of mirror. Its simplicity is consistent with an early eighteenth century date when the current popular style was restrained. The brass candle arm supports at the base probably held glass arms with candle cup ends.

130. Gilded wood, English, circa 1700. The carved crest slides into position above this simple gilded frame and thereby makes this mirror a typical representative of Queen Anne design. A shaped and frequently decorated crest, vertical shape, and plainer base are the basic elements identifying a Queen Anne mirror.

131. Gilded wood, English, circa 1720. The simple molded frame is here embellished with carved detail.

132. Gilded wood, English, circa 1715. The carving details are shallow and restrained on this early example of the Queen Anne style. The reserves in the base originally held candle arms.

133. Silver gilt on wood, English, circa 1700, 32" high, 19" wide. The short-lived fashion for silver gilt was a direct extension of seventeenth century solid silver frames in a less expensive form. Gold leaf, however, was generally preferred and quickly replaced silver leaf as the standard decorative material. This is a lovely mirror with shallow, restrained carving and, in the lower crest, reserves for candle arms.

134. Gilded wood, English, circa 1710. Furniture, mirrors and picture frames gradually became objects where opulence was flaunted and the skill of each carver was displayed. The carved details were undercut to free them of the background and the gilded surface was built up to further accentuate depth. In this early example, a carved head protrudes well beyond the crest while the lower crest supports a concave shell. The mirror frame surface is used as a showcase to display technical carving skill and the variance of light in two and three dimensions. These details are, naturally, a step toward the eventual fantasies which were displayed in the Rococo period of "Chippendale" design. The Queen Anne style can be seen from our perspective as harmonious completeness or a transition period toward flamboyance.

135

136

137

138

135. Gilded wood, English, circa 1730. The pair of birds' heads and grouped feathers are details which became stock motifs in later eighteenth century design. Here they are tamely incorporated into the crest outline, having nothing to do with the shallow leaf carving of the rest of the mirror frame.

136. Gilded wood, English, circa 1730. Here, the outline of the crest is freeing itself from the confines of a rectangular shape as scrolls and a central arch push out the boundaries of the frame.

137. Gilded wood, English, circa 1735. The crest ornaments are somewhat integrated on this mirror where the same elements (head, birds heads, and feathers) all appear emerging from the overall design. The carving of the molded frame is slightly deeper than the previous examples. *Courtesy of M. Turpin Antiques.*

138. Gilded wood, English, attributed to James Moore and John Gumley, circa 1720. This mirror shows a further step in the development of the mirror decoration as the carved head, two birds' heads and three feathers are now integral parts of the overall design, and are joined here by a pair of bearded mens' heads in the crest. The carving overlaps itself and the background design. Here we have the first instance of complete piercing for added depth around the birds' heads. The decoration is becoming unbound from the frame.

John Gumley and James Moore were partners in London between 1714 and 1726 making fine mirrors for the aristocracy. *Courtesy of the Metropolitan Museum of Art, Gift of the Honorable Irwin Untermeyer, 1946.*

139, 140, and 141. Each of gilded wood, English, circa 1710. These three mirrors are cotemporaneous and show variations in the treatment of gilded details, yet a similarity in the frequent use of volutes, scrolled leaves, and carved motifs.

139

140

141

142

143

142, 143, 144. Each of gilded wood, English, circa 1710. Attention to carved detail is rewarded upon close inspection of these mirrors. They each have a variety of symmetrical, petaled crest ornament, but the treatment is vastly different on each. Mirror #142 has a completely removable crest like some of the earlier mirrors already shown.

144

145

146

147

145, 146, 147. Each of gilded wood, English, circa 1710-1720. The progression of crest ornaments on the previous page is continued here with the final design becoming the now-common motif of three feathers which was adopted by the Prince of Wales as an insignia.

72

148. Gilded wood, English, circa 1710-1720,
149. Gilded wood, English, circa 1730, 62½" high,
36½" wide.
150. Gilded wood, English, circa 1730, 61" high, 31"
wide.

   This series is meant to emphasize the development of the carved details from flat to free-standing. The crest and side ornaments gradually became
free from the frame and took on sculptural qualities
not related to their function of supporting the glass
of the mirror. The use of birds' heads was a favorite
device, and each also had candle arms attached to
the lower crests.

151. Gilded wood, English, circa 1740. The most
developed Queen Anne style motifs are combined
in this mirror to make it the epitome of the period.
The central cartouche is completely free of its
structural use, in fact shown emerging from the
tiered shell and expanding outward past the top of
the frame's arch. The flanking scrolls are almost
hidden by the overlapping foliage, and the side
faces are so far removed from usefulness that it
appears that the vines emerging from their mouths
are added just to give them a purpose. These are the
carver's conceit. In the lower crest the shell is nearly
free standing. The light which plays from its varied
textures and surface changes of direction appears as
a contrast to the consistent plane of the mirrored
glass plate.   [See following page]

148

149

150

151

152. Gilded wood, English, circa 1715. This frame has a particularly deep rabbet at the center which suggests that it has always been used with needlework. It compares very closely, however, with contemporary mirror frames, and shows a stage in the progression of pediment details from the arching of the Queen Anne styles to the full, architectural pediment treatment of the Classical which followed. Here the volutes are particularly strong and accented with border gadrooning.

153. Gilded wood, English, circa 1715-1725. The pediment details in this mirror could be seen as advanced for its time, although the egg and dart molding, pierced shell, and drooping swags from rosettes at the volutes are all elements which have appeared separately in previous examples. However, the combination here takes on an architectural design which was further developed into the Constition mirrors later in the century.

152

153

154

155

156

154. **Gilded wood, English, circa 1730.**
155. **Gilded wood, English, circa 1740.** *Courtesy of M. Turpin Antiques.*
156. **Gilded wood and walnut veneer on pine, English, circa 1740.** Each of these three mirrors possesses details which were further developed into the Constitution style mirrors a few decades later. In #154 and 155, the background punching sets off the carved decoration just as the solid wood veneer on #156 and later examples contrast with the gilded details. The pediment elements, swags and molding treatments all were precocious Queen Anne motifs which contributed to even more free-standing designs to come. The quality of the carving in each is superb.

157

158

Transitional

157. Gilded pine, English, circa 1743. It is likely that this mirror was designed by William Kent for Frederick Louis, Prince of Wales, about 1743. (Frederick Louis was the son of King George II and father of King George III.) The design has roots in Daniel Marot's designs of the seventeenth century (see picture 52) as they were reinterpreted by the mid-eighteenth century designers. William Kent was a leader in this style of design. His work is generally heavy and architectural.

158. Gilded wood, English, circa 1740, 48" high, 27" wide. The treatment of this carving is heavy overall, but with carefully planned piercings meant to lighten the design. The mask in the lower crest is a carry-over from Renaissance designs.

159. Gilded wood, English, circa 1730-35. 34" high, 28" wide. Here is another example of the gradual loosening of carved detail where piercings lighten the crest design. The trend during the 1740's was toward lighter, and more open designs, and from rounded to taller rectangular shapes.

159

160

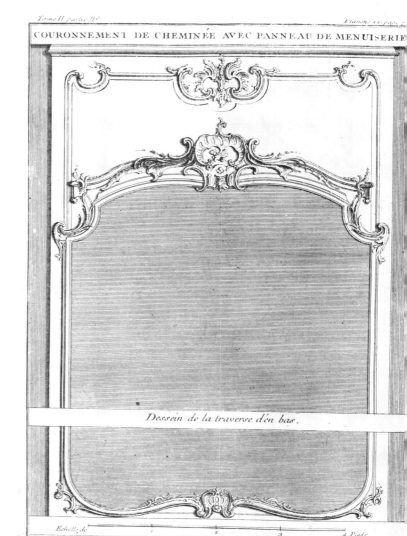

COURONNEMENT DE CHEMINÉE AVEC PANNEAU DE MENUISERIE

Dessein de la traverse d'en bas.

Echelle de   1   2   3   4 Piede

161

160. Gilded wood, English, circa 1730. French design sources are evident in this mirror and matching marble top pier table. Coordinated furniture was frequently designed for large royal homes, with details matching architectural motifs.
161. Design for an overmantle mirror by J.F. Blondel from "De la Distribution des Maisons de Plaisance." Paris, 1738. French mirrors at this period were frequently part of the room paneling. *Courtesy of the Victoria and Albert Museum.*
162. Room from Palais Paar, Vienna. French designs influenced all of the Rococo craftsmen in Europe, this room being but one of the countless chambers with mirrors incorporated into the paneling. *Courtesy of The Metropolitan Museum of Art.*

162

163

164

165

163. Gilded wood, English, circa 1730. William Kent's heavy architectural style is seen in the columns supporting this enormous and fanciful mirror.

164. Painted oak, French, early eighteenth century. A panel from a French room is carved for both painted and mirrored glass panels. *Courtesy of The Metropolitan Museum of Art, Gift of J. Pierpont Morgan, 1906.*

165. Gilded wood, English, made by Benjamin Goodison in 1732. This is one of three glass sconces in carved and gilt frames with two wrought arms each made by Benjamin Goodison for the Prince of Wales in 1732-3, which are at Hampton Court Palace. One candle cup is now missing. The rough texture in the frame is gilded sand. *Reproduced by gracious permission of Her Majesty the Queen.*

166

166. Gilded wood, English, circa 1730. The architectural members of this design, such as the scrolled pediment, dentiled molding, convex molding and crossetted top corners, link this design to the strictly ordered system of details which determined the Georgian period of decoration. Here, every surface is carved. The overall feeling is one of mass and solid material in the frame contrasting with the void of the glass. *Courtesy of the Victoria and Albert Museum.*

167. Pencil drawing by John Vardy, English, circa 1745. Architect John Vardy drew this pier table and mirror. Comparisons can be made with it and the designs of William Kent, his close associate. The same classical motifs were used again at the end of the eighteenth century by Robert Adam but interpreted in a lighter fashion. *Courtesy of the Victoria and Albert Museum.*

168. Gilded wood, English, circa 1730. The heavy architectural design of this mirror is lightened by the leafy swags in the top and bottom crest, and the mirrored area is completely dominated by the superimposed clock, barometer, and thermometer. The carving is light, free standing, and crisp, but the overall design is still dominated by the heavy frame with gilded sand as a texture contrast. *Courtesy of the Victoria and Albert Museum.*

167

168

169

169, 170, 171. Carved wood by Mathias Lock, circa 1743, 103" high, 50" wide. This mirror and table have survived with an accompanying drawing inscribed 'A Large Sconce for the Tapestrey Roome' and a bill for £36 5s (of which the carving accounted for £34 10s.) They were probably made for Hinton House, Hinton Street, George, Somerset in 1743. As no reference in the bill is made to gilding, and the present bronze decoration has been determined to date from the early nineteenth century (J.F. Howard, *"Furniture by Mathias Lock for Hinton House," Connaisseur,* vol. CXLVI, December, 1960. pages 284-6). We conclude that it was not colored as now originally.

The carving is ambitious, relating to a hunt with costumed heads, rabbit, dogs, dolphins, pipes, birds and lion mask all joined by C-scrolls, leafage and fruit garlands. The table particularly has an overall French feeling. Seen as a transition piece between the heavy Queen Anne and light "Georgian" to follow, this group is a masterpiece of integrated details and superb carving. *Courtesy of the Victoria and Albert Museum.*

170

171

172

172. Gilded wood and etched glass, Venice or Genoa, mid-eighteenth century, about 96" high. This mirror displays decorative details which have their roots in the Italian Renaissance. The etching of patterned, assymmetrical panels linked by scrolled floral vines, the large cartouch in the crest, and swelled arch at the top are Italian designs which were borrowed first by the French, and then by the English designers in the eighteenth century. Design, as a fluid continuum is exhibited through this mirror and the examples to follow. Portions of the decorations will be seen reinterpreted on later mirrors through the "Chippendale", & "Adam", periods. *Courtesy of the Victoria and Albert Museum.*

173. Carved wood, French or Italian, circa 1730. The exotic dragon, lion mask, and pair of mermaids on each mirror are carved "in mirror image" on this pair. The overall upward swelling shape is a European style which was not often attempted in England or America. There, symmetry and regular shape were usual. Baroque C-scrolls link each motif with leafage and cabochons. In England, these details became exploited by designers such as Thomas Chippendale who popularized and influenced English and American design later in the eighteenth century. *The Wallace Collection.*

174. Oil on canvas painting "Nobleman kissing a Lady's hand" by Pietro Longhi, Italian, mid-eighteenth century. The mirror in this picture has a more restrained frame, but is in the tradition of the previous ones in this group. The upward bulging shape is present and scrolls surround the glass. Pietro Longhi (1702-1785) painted fashionable interiors from his native Venice, the source of so much of eighteenth century design. *Reproduced by courtesy of the Trustees, The National Gallery of Art, London.*

173

175

176

175. Gilded wood, English, circa 1750. The symmetrical shape and use of C-scrolls to link the oclumns are Baroque features of this mirror. New are the use of a cabochon and icicles and the intertwined candle arms—all features which were developed in the emerging "Chippendale" or "late Georgian" style in England.

176. Gilded wood and engraved glass, Venetian, early eighteenth century, 42" high, 28½" wide. The overall symmetrical shape of this mirror is a Baroque feature, yet the assymmetrical crest carving is an early Rococo feature. While the carving of the frame is mediocre, the glass is beautifully etched in the tradition of Italian decoration. *Courtesy of the Victoria and Albert Museum.*

177. Painted and gilded wood, Venetian, mid-eighteenth century, 33" high, 25" wide, 11" deep. The use of painted wood surfaces was prevalent in Venice. The dressing mirror and its stand have no straight lines; the eye is made to move all around the space. While the frame is symmetrically balanced, it is a heavy molding made to seem less heavy by its painted decoration. The supports, although thick, are curved on each plane and tapering, exhibiting the style passing from exuberant Baroque toward flamboyant Rococo. *Courtesy of the Victoria and Albert Museum.*

178

83

178. Gilded wood, South German, Second quarter of the eighteenth century, 30" high, 32" wide, 7" deep. The three cherubs are vestiges of Baroque design incorporated into this Rococo, assymmetrical, ever-moving design. It is the height of German Rococo. The lion's head and dolphin head in the base add weight to anchor the dressing mirror which otherwise, in design, has no focus-but floats as a moving cloud. This last feature typifies the emerging Rococo style as cloud-like.

179. Silver, German, early eighteenth century. These are parts of a toilet set of over fifty pieces made in Augsburg. It displays Baroque details in its symmetry and C-scrolls, but emerging Rococo features in the assymmetrical crest and base cartouches. The crest cartouch supports an engraved peacock, while the base cartouch shows a parrot in relief. The interesting pattern of the glass beveling softens the change from the frame to the glass. *The Wallace Collection.*

179

ROCOCO

Architectural

180. Gilded wood, English, circa 1740. The architectural elements of this mirror, including the closed, straight pediment, frieze, keystone, crossetted top corners, and scrolled base corners, are derived from Palladian building designs and the work of such interior designers as William Kent in the early eighteenth century. The quality of the raised carving is superb. Among the "new" details are the side bell flower swags and leaf garlands, which were used in profusion at the end of the century.

This mirror has the architectural shape, and side decorations which identify a style of mirror known as "Constitution"—which will be seen in many of its emering forms in this chapter. The origin of the term "Constitution" is not known, yet its use to describe this style is widely understood.

181. Probably walnut, English, circa 1740. This mirror frame is absolutely architectural with no hint of ever having further ornament. The broken arch and molding rising into the frieze, crossetted corners and molded inner edge are all details which were modified and embellished in the architectural or "Constitution" mirrors which evolved. *Courtesy of the Victoria and Albert Museum.*

182. Gilded wood, English, circa 1740, 46″ high, 26″ wide. The straight broken pediment of this architectural mirror and the decoraton confined to the area within the frame suggest an early date. The blank areas in the bottom scrolls are designed for the application of sconce arms. *Courtesy of the Victoria and Albert Museum.*

183. Walnut veneer and gilded wood, English, circa 1735, 66″ high, 34″ wide. The straight broken pediment and heavily carved drapery swags suggest the designs of William Kent and propose an early date for this architectural mirror. *Courtesy of Alfred Bullard, Inc.*

184. Gilded wood, English, circa 1730. The scrolled pediment appears on this mirror with the coat of arms of Sir William Bowes of Stratham Castle, Durham. The frieze bears a carved mask and bundled feathers. In the lower corners, candle arm supports are in place, and in the lower crest, a pair of bird's heads and a small face appear. These are quaint details which reappear throughout this period and into the Rococo style. *Courtesy of The Metropolitan Museum of Art.*

185. Gilded wood, English, circa 1735, 60″ high, 30″ wide. The architectural features of this mirror are accompanied by a large central cartouch, side swags, tiny bird heads and small scrolls in the lower corners. These features are rudimentary to the "Constitution" style.

180

181

5906

182

183

184

185

186

187

186. Gilded carved wood, English, circa 1740. The architectural details of Constitution mirrors are all present in this early mirror: the pediment, frieze, lower scrolls and crest. Here, however, the entire surface is gilded. The later development was for the flat areas to be veneered with mahogany or walnut.

187. Gilded carved wood, English, circa 1740. This architectural mirror has an early style swag fitting closely to the side moldings. The carving is carefully balanced and beautifully executed.

188. Carved wood, English, circa 1730, 65" high, 27" wide. An early gilded architectural mirror has a sand covered background which is lightly gessoed and gilded. On this example, the sconce arms are possible later additions, the originals probably having been attached at the oval rosettes in the lower crest. *Courtesy of the Tryon Palace Restoration, New Bern, North Carolina.*

188

189. Walnut veneer and gilded wood, English, circa 1730, 85" high, 38" wide. Experiments in background treatment to set off the gilded carving (such as sand in the previous mirror) resulted in the widespread use of wood veneers, this one a particularly mottled burled walnut. This treatment became the standard for "Constitution" mirrors. *Courtesy of Alfred Bullard, Inc.*

190. Mahogany veneer and gilded wood, English, circa 1750. The long proportions of this mirror are more common with Baroque (or "Queen Anne") mirrors than with "Constitution" architectural mirrors of which it is a fine example.

191

192

193

194

195

196

191. Walnut veneer on pine and gilded wood, English, circa 1730-1740, 70″ high, 35″ wide.
192. Walnut veneer and gilded wood, English, circa 1750. *Courtesy of Alfred Bullard, Inc.*
193. Walnut veneer and gilded wood, English, circa 1740, 57″ high, 30″ wide. *Courtesy of Alfred Bullard, Inc.*
194. Walnut veneer on pine and gilded wood, English, circa 1740, 45″ high, 24¾″ wide. *Courtesy of the Victoria and Albert Museum.*

Walnut veneer seems to have been used before mahogany veneer on these architectural mirrors. A close examination of the gilded details shows considerable variations among the mirrors. The cartouches and frieze decorations are most apparent as they vary in size and design so to completely change the overall appearance of each mirror. The scrolled lower corners are variously shaped and ornamented as well.

195. Walnut veneer and gilded wood, American or English, circa 1740. *Courtesy of David Stockwell, Inc.*
196. Walnut veneer and gilded wood, American or English, 1740-1750, 81″ high, 23″ wide. *Diplomatic Reception Rooms, United States Department of State.*

The appearance of gilded eagles in the crest adds an assymmetrical finishing detail to these architectural mirrors. That, and long histories of use in American families suggest that these may be of American origin. Proof of that origin, however, is inconclusive. At their date of manufacture, it is most likely that they were of English origin.

197. Walnut veneer and gilded wood, English, circa 1740, 54" high, 31½" wide. *Courtesy of Alfred Bullard, Inc.*
198. Walnut veneer and gilded wood, English, circa 1740. *Courtesy of Alfred Bullard, Inc.*
199. Walnut veneer and gilded wood, English, circa 1740, 55½" high, 27¾" wide.

Many of the ornamental details on these architectural mirrors were used increasingly in the following Rococo period.

197

198

199

201

200

200. Mahogany veneer on pine and gilded wood, American or English, circa 1765, 80″ high, 34″ wide. This mirror has a history in Albany, New York from about 1774 when it was in the estate of Sir William Johnson (1715-1774), Colonial Commissioner of Indian Affairs. Later, it descended from John Tayler (1742-1829) Lieutenant Governor of New York, one of the wealthiest men of his time. *Diplomatic Reception Rooms, United States Department of State.*

201. Walnut veneer and gilded wood, English, circa 1750, 66″ high, 34″ wide. The basket finial and swags descending only part of the way down the sides are features which will be seen again on mirrors in the Classical style at the end of the century. This mirror also has incised and gilded decoration in the crest which is unusual for this group of architectural mirrors and shows a detail which is tending toward the elaborate, flamboyant Rococo style which followed. *Courtesy of the Victoria and Albert Museum.*

202

203

202. Mahogany veneer and gilded wood, English or American, circa 1750. The tall, divided, shaped, and cut panels of glass in this mirror recall the earlier mirrors in the Queen Anne period. The use of profuse intaglio carved and gilded details, however, resemble the fancy Rococo mirrors popular twenty years later. The use of all this decoration makes this mirror a spectacular blend of the architectural and Rococo motifs. *Courtesy of David Stockwell, Inc.*

203. Mahogany veneer and gilded wood, English, circa 1760. Both mirrors are fine examples of the architectural style, yet the various details of decoration produce quite different final results, one much lighter in effect than the other.

204. Walnut veneer and gilded wood, English, circa 1740, 55½" high, 22" wide. *Courtesy of David Stockwell, Inc.*

205. Walnut veneer and gilded wood, English, circa 1740, 55½" high, 22" wide. *Courtesy of David Stockwell, Inc.*

204

205

206

207

206. Mahogany veneer and gilded wood, English or American, circa 1770. The Phoenix finial is original to this mirror. *Diplomatic Recption Rooms. United States Department of State.*

207. Mahogany veneer and gilded wood, English or American, circa 1750-1775, 76" high, 30" wide. This mirror has a history from New York and bears an original eagle finial. *Diplomatic Reception Rooms. United States Department of State.*

208. Mahogany veneer and gilded wood, English, circa 1760. The upper portion of this mirror has ornate gilded details which lighten the overall architectural feeling of the lower section. *Courtesy of Herbert Schiffer Antiques.*

209. Walnut veneer and gilded wood, English, circa 1740. *Courtesy of Alfred Bullard, Inc.*

210. Walnut veneer and gilded pine, English, circa 1750. *Courtesy of Alfred Bullard, Inc.*

208

209

210

211

212

213

214

215

216

217

218

211. **Mahogany veneer and gilded wood, English or American, circa 1750.** *Courtesy of Herbert Schiffer Antiques.*
212. **Mahogany veneer and gilded wood, English, circa 1760.** *Courtesy of Herbert Schiffer Antiques.*
213. **Walnut veneer and gilded wood, English, circa 1740, 55" high, 26" wide.**
214. **Mahogany veneer and gilded wood, English, circa 1750.** *Courtesy of Alfred Bullard, Inc.*
Since Americans like to relate to their national symbol, the eagle, mirrors with eagle finials are especially popular in American collections.
215. **Mahogany veneer and carved wood, English, circa 1760.**
216. **Walnut veneer and gilded wood, English, circa 1740.** *Courtesy of Herbert Schiffer Antiques.*

217. **Mahogany veneer and gilded wood, English, circa 1760, 55½" high, 27¾" wide.** *Courtesy of Herbert Schiffer Antiques.*
218. **Mahogany veneer and gilded wood, English, circa 1740-1750.** *Courtesy of Alfred Bullard, Inc.*

219

220

221

222

223

224

219. Mahogany veneer and gilded pine, English, circa 1760. *Courtesy of Peter and Nancy Schiffer.*

220. Walnut veneer and gilded wood, English, circa 1750. There is increasingly more open space in the carved side garland swags on this architectural mirror than mirrors of apparently earlier date. The trend toward lighter details is consistent with the artistic movements of the eighteenth century. Piercings increase and carving becomes more refined and under cut as the century progresses. *Courtesy of Alfred Bullard, Inc.*

221. Mahogany veneer and gilded pine, American signed by Simon Deming of Wethersfield, Connecticut, circa 1790 to 1800, 57" high, 26" wide. The late date of this pair of mirrors provide an example of America's cultural lag in style development. The floral urns are a very late type of finial decoration for architectural mirrors, being more commonly associated with Classical style mirrors. The outside moldings are notably free of carving, which also suggest a late date.

Simon Deming worked in Wethersfield, Connecticut before forming the partnership of Wells and Deming in New York. *Courtesy of Israel Sack, Inc., New York City.*

222. Mahogany veneer, gilded wood, English, circa 1760 to 1770. A lighter overall feeling is achieved on this mirror by the simple, uncarved outer moldings, pierced leaf swags, and widely spaced volutes. *Courtesy of Herbert Schiffer Antiques.*

223. Mahogany veneer and gilded wood, English, circa 1750. *Courtesy Philadelphia Museum of Art. Purchased from a gift from the Trust under the deed of Rodman Barker to the City of Philadelphia and from the Museum Funds.*

224. Walnut veneer, English, circa 1750, 53" high, 27" wide.

These mirrors include phoenix finials and open space among the swags to give a relatively late date of origin, yet #224 especially has an early feeling in its proportions and beautiful quality sconce arms and attachment plates. *Courtesy of Herbert Schiffer Antiques.*

Gilded

225. Gilded wood, English, circa 1760, 91" high, 38" wide. The heavy carved figures are a carryover from the Baroque styles, but here the light carving with cut-out areas signal the emerging Rococo style. The tendency at the second half of the eighteenth century was toward light, assymmetrical design with increasing imagination. As the Rococo elements gradually dominated interior furnishings, the older "Queen Anne" or Baroque style persisted, but with less enthusiasm than the new Rococo.

This mirror may have been made by Samuel Norman and James Whittle who were partners for a time around 1759 in London. *Courtesy of the Victoria and Albert Museum.*

225a. Gilded wood, English, circa 1735-1740, 53½" high, 32" wide. One of the delights of the Rococo style is deception and the beginnings of it are seen on this mirror. Leaves overlap the frame and the pierced carvings in the crest, and the frame is concealed here by a c-scroll and there composed of c-scrolls. The in and out play became a game between the artist and the observer.

225

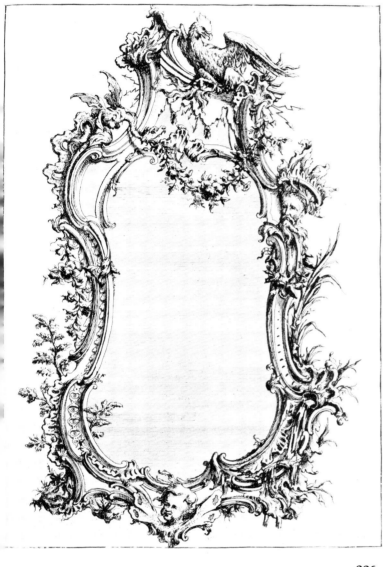

226

226. Drawing by Matthias Lock, from "Six Sconces", London, 1744. This flamboyant drawing is an interpretation of French Rococo designs of this period. The details are whimisical here, and probably never inteded to be copied in all this profusion; but as a sample of the different motifs that could be used one or two at a time to ornament a mirror. *Courtesy of the Victoria and Albert Museum.*

226a. Gilded wood, English, circa 1760. The prominent designers of this period were each playing with the freedom the new Rococo style allowed them. William and John Linnell, William Ince, John Mayhew, Thomas Johnson, Thomas Chippendale and other of the designers used birds, scrolls, leaves and off-center designs to explore space and visual deception. *Courtesy of the Victoria and Albert Museum.*

227

228

229

230

231

232

227. Gilded wood, English, circa 1750. The use of all gilded frames was begun in Europe, especially in France where ornate furnishings became a source of competition among the aristocracy. The designs and spirit were quickly transported to England where designers published books on drawings to publicize the new, freer, whimsical, Rococo style.
228. Gilded wood, English, circa 1750.
229. Gilded wood, English, circa 1750.
230. Gilded wood, English, circa 1760, 53″ high, 31″ wide.
The transplanted French designs developed into their own style in England, with elements such as the cabochon-shaped solids and open areas becoming integral motifs. The variations in detail are apparent, but overall lightness and superb carving techniques are shared by the members of this group of mirrors.

231. Gilded wood, English, circa 1755, 65″ high, 47″ wide. The lightness of carving detail was not technically possible to achieve on mahogany or other hard woods that were used in the earlier period. Only after white pine was imported in quantity to England from America could a popular style be supported with ornately carved details. The pine is easily carved, then prepared with gesso and gilded. This mirror is a brilliant feat of design and carving skill. *Reproduced by Gracious Permission of Her Majesty the Queen.*
232. Gilded wood, English, circa 1755, 45″ high, 22″ wide. The design preference was ever more open.

233

234

235

233. Drawing by Peter Babel in 1752, engraved by G. Bickham, English. Peter Babel interpreted the French Rococo style and growing fascination with oriental motifs in this drawing for a mirror in 1752. A Chinese man's face, pagoda, dragons, birds and flowers are all linked by c-scrolls. The assymmetry and imagination are a delight here. The clouds at the crest are sheer imagination which is possible on paper, but does not reproduce easily in carving. *Courtesy of the Victoria and Albert Museum.*

234. Drawing "Designs for pier glasses", No. 173, by Thomas Chippendale, 1754. This is the drawing from which an engraving was made for "The Gentleman and Cabinet Maker's Director" by Thomas Chippendale, published first in 1754. Chippendale worked in the prevalent Rococo style of his time creating imaginery space with pastoral themes and was influenced by French and Chinese designs. His book was enormously influential in Europe and America for the entire rest of the eighteenth century.

235. Gilded wood, probably Italian, circa 1740. This Italian Rococo mirror has certain elements which appeared in French, English, American and subsequent interpretations of the Rococo style. The c-scrolls and assymmetrical shells are particularly important on this mirror.

236

237

238

236, 237, and 238. Gilded wood, English, circa 1750. These large mirrors in the early Rococo style are gilded today, but ones of this type are also known to have been painted greenish-grey-white originally. Those which have survived with the painted surface are very carefully carved and of the best quality. Later, some of them were most probably gilded. The use of glass behind the carving has precedence in the early Baroque mirrors with mirrored frames. In the Rococo style, the use of varying levels of reflection and carving is part of the teasing nature of the design.

239

240

239. Gilded wood, English, circa 1760, 48″ high, 25″ wide.
240. Gilded wood, English, circa 1760. *Courtesy of M. Turpin, Inc.*
241. Gilded wood, English, circa 1760, 40″ high, 20″ wide.
These three mirrors may have derived from the same drawing, as many details are similar. Different carvers, however, are responsible for the variations in details such as the use of cabochons in the upper corner ruffles of #240 and icicles, and the crest finial of #241.

242 and 243. Gilded wood, English, circa 1760. The slight variations of Rococo design make delightful comparisons of these elegant mirrors.
244. Gilded wood, English, circa 1750, 73″ high, 27″ wide. The tall proportions of this Rococo mirror resemble two-piece Baroque mirrors of the Queen Anne period. The carving on this is excellent, perhaps by Thomas Johnson or a contemporary master carver (see also #301).

241

242

243

244

245

246

245. Gilded wood, English, circa 1750. The open piercing of this mirror frame is typical of the well developed Rococo style. The galleried platform on top may have supported a carved figure originally.

246. Painted pine, attributed to James Reynolds in Philadelphia, in 1772, 78" high, 42" wide. This is one of only a few known American-made Rococo mirrors surviving today. It is one of a pair (the other reflected) still hanging in its original house "Cliveden" in Philadelphia, the home of Benjamin Chew and his descendents. The surface is painted white, as it always has been. No trace of gilding exists. James Reynolds adapted a design from one of the fashionable Rococo style design books circulating at the time. The frame was made apparently for pieces of glass already in Philadelphia from a Queen Anne style mirror. The glass determined the size and 2-panel shape of the mirror, and required the double set of c-scrolls at the top corners to conceal the curved, beveled glass there. They are quite remarkable by any standard, and stand nearly unique in American manufacture.

247

248

249

The related American Rococo mirrors are a pair of sconces of the same white surface also attributed to James Reynolds and also hanging at Cliveden. Another mirror known as the Fisher Family mirror is painted white and partly gilded. A final mirror of the type shown as #247 in this book, is painted white and partly gilded and was made by James Reynolds for John Cadwalader of Philadelphia; it now is in the collection of the Henry Francis du Pont Winterthur Museum. (See "Cliveden and Its Philadelphia-Chippendale Furniture: A Documented History," by Raymond V. Shepherd, Jr. *The American Art Journal,* Volume VIII, Number 2, November, 1976.)

247. Painted white and gilded pine, made by James Reynolds in Philadelphia, circa 1770, 55½" high, 28¼" wide. This is one of the few known American made Rococo style mirrors remaining. Similar to the mirror #246, this one was made in Philadelphia by James Reynolds for John Cadawalder in 1770. It has superb carving. *Courtesy, The Henry Francis du Pont Winterthur Museum.*
248 and 249, Gilded wood, English, circa 1760.

250

251

252

253

254

255

250. Gilded wood, English, circa 1760. *Courtesy of M. Turpin Antiques.*
251. Gilded wood, English, circa 1760.
252. Gilded wood, English, circa 1780, 49" high, 25" wide.
253. Gilded wood, English, circa 1760. *Courtesy of M. Turpin Antiques.*

The continued use of open space in the designs creates a lacy overall feeling. In #253, bell-flower garlands hang down the sides of the frame and in the crest. This motif was popularized by Robert Adam toward the end of the eighteenth century as tastes evolved toward ever-thinner Neo Classical designs.

254. Gilded wood, English, circa 1760, 60" high, 30" wide. The popular, mythical phoenix bird became a finial ornament on Rococo mirrors. The pheonix, not to be confused with the eagle, has a crest, is scrawny, long-neck, and usually is shown somewhat contorted, as his rise from fire ashes in mythology precludes.
255. Gilded wood, English, circa 1750-1755. The large cabochon shaped carvings in the ruffles along the top frame of this mirror were a motif used to indicate overlapping space. In American design they are sometimes called "peanuts" and are found on highly carved chairs and chests of this Rococo style.
256. Gilded wood, English, circa 1760, 55" high, 21" wide. The carving is rather stiff in this mirror lacking the overlapping of other contemporary examples. The bird looks uncomfortable.

256

257

258

259

260

261

257. Gilded wood, English, circa 1770, 62″ high, 28″ wide. This is a light Rococo design with icicles, side columns, and quite a fanciful phoenix.

258. Painted and gilded wood, English or Continental, circa 1750, 105″ high, 62″ wide. The Chinese influence on Rococo design was evident on the Continent before England or America. This fantastic overmantle mirror may have been made in Europe or England. Paint, lacquer, gilding on a ruby lacquered ground and normal gilding all are parts of the surface decoration. The design is extremely ambitious and seems to be without support. The quality of the carving is extremely delicate, perhaps surpassed only by that of the reflected mirror (#262.) They make a fascinating picture. Another view of these can be seen reflected in picture #260. *Courtesy of the Victoria and Albert Museum.*

259. Gilded wood, English, circa 1760, 48″ high, 54″ wide. This Rococo overmantle mirror was designed "in the Chinese taste" to hold small Chinese porcelain objects, as shown. As porcelain objects became admired and collected, their display became of concern to interior designers. This use of

a mirror and frame was a novel solution linked with room paneling and separate shelf displays. *Courtesy of the Victoria and Albert Museum.*

260. Gilded wood, English, circa 1760. The designers went wild creating imaginery space. Here bells hang from the pagoda, curving columns support c-scrolls as fantastic phoenixes rest on the side branches.

Another delight of this picture is the reflection of mirror #258. *Courtesy of the Victoria and Albert Museum.*

261. Gilded wood and eglomise over mirrored glass, English, circa 1750. The frame alone is a fine example of Chinese style Rococo design, (Chinoiserie) with peaked roof motifs, birds, bells, and railing pattern in the crest. Combined with the eglomise reverse painted picture on glass of Oriental figures in a Chinese style landscape, the mirror becomes an excuse for ever more ornate and exotic decoration. This is a superb example of eighteenth century Rococo design.

262

263

264

265

262. Gilded wood, English, circa 1760, 133" high, 38" wide. This Anglo-Chinese confection is one of a pair over eleven feet tall with combined English and Chinese elements. It is about as complicated a design as has been carved. *Courtesy of the Victoria and Albert Museum.*

263. Gilded wood, English, circa 1760, 84" high, 54" wide. The three twisting candle arms emerge from the lower crest like branches of the gnarled tree trunk. Overlapping foliage and c-scrolls, birds and flowers make an elegant and enormous sconce. Pairs of sconces and even larger matching pier mirrors were made for gigantic palace rooms. *Courtesy of the Victoria and Albert Museum.*

264. Gilded wood, English, circa 1760. Since pastoral representation became an important aspect of Rococo design, fables such as those told by Aesop became sources for decoration at this period—as this one of the fox and the crane depicts. A mirrored panel belongs in the rectangle behind the crane.

265. Gilded wood, English, circa 1760. The assymmetrical design of each sconce is balanced by the mirror-image of the other member of this pair. The overall shape resembles Chinese pagodas and the candle arms are again branches, as in mirror #263.

266. Painted and gilded wood, English, circa 1760. Chinese figures in fantastic costumes were a Rococo exotic affectation. This sconce has two cups for candles. It was made for Ditchley House in England.

266

267

268

269

270

271

272

267. Gilded wood, English, circa 1760, 55¾" high, 32" wide. *Courtesy of Alfred Bullard, Inc.*
268. Gilded wood, English, circa 1770.
269. Gilded wood, English, circa 1770.
Chinese pagodas are variously combined with Rococo design including small country houses. It is curious that in an imaginery space such as these mirrors display, it is perfectly possible and characteristic to find pastoral English details incorporated with exotic Chinese and completely imaginery motifs.
270. Gilded wood, English, circa 1760. The French taste for pastoral scenes is carried over to this mirror design. Chinese style gates are combined with grazing sheep in an assymmetrical shape.
271. Gilded wood, English, circa 1760, 42" high, 19½" wide. It is sometimes hard to differentiate between the support and the decoration on these later Rococo mirrors. This mirror probably had a candle supporting arm originally.
272. Gilded wood, English, circa 1760. The carving is magnificent on this sconce where many of the Rococo details are combined. Two candle arms spiral out of the lower design.
273. Gilded wood, Irish, circa 1760. Although needed repairs are evident in this picture, the overall shape and carving display good Rococo design.

273

274. Gilded wood frame, English, circa 1760 around Chinese eglomise mirrored glass. The frame is a good example of mid-eighteenth century design for painted portraits and landscapes. Here it is supporting a fine Chinese-painted eglomise panel with mirrored background which was sent from England to China for decorating.

275. Gilded wood frame, English, circa 1760 around Chinese painted glass. When the European taste for Chinese art was at its zenith in the mid-eighteenth century, glass panels were sent from England to China for decoration. Here a fine English frame surrounds the Chinese-decorated panel with mirrored background. Comparisons of this bird and floral design can be made with the English Chinoiserie frames in this section.

276. Gilded wood frame, English, circa 1760 around Chinese painted glass. The Chinoiserie design is quite architectural with restraint in the decorative aspects of this mirror.

277. Gilded wood frame, English, circa 1760 around Chinese painted glass.
278. Gilded wood frame, English, circa 1760 around Chinese painted glass, 46″ high, 31¾″ wide. *Courtesy Tryon Palace Restoration, New Bern, North Carolina. Gift of Mrs. James Edwin Latham.*
279. Gilded wood frame, English, circa 1760 around Chinese painted glass.

274

275

276

277

278

279

Overmantle

281

280

280. Drawing, by Abraham Swan in 1745, English. Abraham Swan published this design for an overmantle mirror in his "Gentlemen's or Builder's Companion" in 1745. He combined Baroque architectural design with the emerging lighter Rococo style. *Courtesy of the Victoria and Albert Museum.*

281. Carved wood, English, circa 1750-1755. This entire fireplace surround was designed and executed as one continuous piece. The carved mantle and mirror frame are integrated by the c-scrolls and floral details in both members.

282. Carved wood, English, circa 1760. Small ledges are carved into the mirror decoration and are intended to hold small pieces of Chinese porcelain. The gothic influence on Rococo design is quite evident here. The carving of the mantle piece is quite outstanding.

283. Painted wood, English, circa 1750. Rococo exuberance is displayed by this mantle piece and mirror designed as one piece for Winchester House, Putney, England. The carving is painted white, not gilded, as a few choice examples seen here have been. One's eyes are not stationery in front of this piece. *Courtesy of the Victoria and Albert Museum.*

282

283

284. Gilded wood and oil on canvas painting, English, circa 1760-1770, 69½" wide, 81" high. The Rococo love of the exotic is no better illustrated than here where an imaginery Oriental landscape painting is framed with a mirrored glass panel. The frame includes the customary birds, pagoda and foliage that became standard motifs of this period. The painting is derived from an engraving by Johan Bernhard Fischer von Erlach which appeared in his book *Entwurff einer Historischen Architectur* (Leipsig, 1725). The scene itself is an imaginery one including an approximation of a view of the Siam Harbor (now Bangkok) with a Chinese Nanking tiered pagoda. The truth never hindered European imagination. [*Correspondence in the files of Tryon Palace Restoration*] *Courtesy Tryon Palace Restoration, New Bern, North Carolina.*

285. Gilded wood, English, circa 1750, 50" long, 30" high. Three small pieces of glass precluded the vertical dividers, and here they are integrated into the design of the frame.

286. Walnut veneer and gilded wood, and oil on canvas painting, English, circa 1745. This overmantle mirror is probably a unique design. The oil painting depicts an imaginery garden with Classical ruin in the English countryside. Below, the mirrored glass panel was very valuable. These two were framed together to fill a specific space above a fire place. The frame has been designed carefully to set off both the mirror and the painting equally, with only a few architectural and ornamental elements of its own to give it unity. The use of walnut veneer relates this frame to the Architectural frames grouped together in this book. *Courtesy of Alfred Bullard, Inc.*

287. Gilded wood, English, circa 1750.

288. Gilded wood, English, circa 1750, 64" long, 32" high.

284

285

286

287

288

291. Gilded wood, English, circa 1760. Two pair of sconce arms unwind from the base tree design on this overmantle mirror. The French and then English delight in imaginery pastoral scenes is brought into this design by the pair of sheep, monkeys, bird (needing restoration) and gnarled, broken tree trunks. This unlikely group of characters is brought together in this design and made believable only because the whole design is more clever than reality.

292. Carved wood, English, circa 1770, 37″ high, 53½″ wide. The very light design and oval shape indicate a step away from the massive and complicated Rococo designs toward symmetry and order. Small Classical urns are resting on the lower plinths, as this is a step closer to the Classical style which gradually overtook Rococo by the end of the eighteenth century.

289

289. Gilded wood, English, circa 1760, 50″ high, 52″ wide. Four phoenix birds in this airy, Rococo mirror design are just teasing details when compared with the massive crest carving of the mythical tale of Ganymede and the Eagle. The story of Ganymede was also depicted by Thomas Johnson in his design book *Twelve Girandoles* of 1758, (plate 36) and again in his book *One Hundred and Fifty New Designs*. This mirror creates a lovely, imaginery space.

290. Gilded wood, English, circa 1750-55. Among Rococo overmantle mirrors, the use of an oval shape is unusual. The scrolls and foliage are particularly well carved on this mirror, and the phoenix is extremely animated.

290

291

292

293

294

295

Oval

296 and 297. Gilded wood, English, circa 1760-1770. These two oval mirrors stand as examples of the transitional style from Rococo to Neo-Classic as they utilize fine quality carved foliage and c-scrolls with bell flowers and a symmetrical, open design. The carving on #296 is particularly crisp, perhaps indicating that it was originally left ungilded as a few of the most carefully carved mirrors were.

296

293. Gilded wood, English, circa 1760-1770, 37" high, 60" wide. The symmetry, light frame, and rope of wheat husks or bell flowers were used in the further development of the style into Neo Classical.
294. Gilded wood, English, circa 1760-1770. This Rococo overmantle mirror shows the late development of the style in its symmetry, light tracery design and use of an oval panel. The two feet at the base support the mirror above the mantle shelf.
295. Gilded wood, English, circa 1760. The triple panels of this overmantle mirror resemble the gothic influences of the Rococo style, while the side panels are capped by Chinese inspired pagoda tops. Such a wide group of influences worked together to create this style. The carving is light and controlled.

298

298. Gilded wood, American, circa 1770. This delicate mirror was probably made in Philadelphia. The carving is of good quality. It may be related to the fine white painted mirrors by James Reynolds shown as numbers 246 and 247 in this book.

299. Gilded wood, English, circa 1760. This small round mirror has a restrained frame to set off the fantasy going on in the crest. The carved squirrel huddles under a pagoda of curling leaves. Such nonsense and delight.

300. Gilded wood, English, attributed to John Linnell, circa 1765-1770, 78″ high, 40″ wide. John Linnell worked for some of the wealthiest clients in England during the second half of the eighteenth century in the currently fashionable tastes, including early Baroque, Neo-Classical, and the Rococo style. His mirrors are large and characteristically light in detail. The pair of Classical urns and basket in the crest are less Baroque than Neo-Classic, suggesting the trend in the later Rococo period toward the latter and a fairly late date for this mirror.

299

300

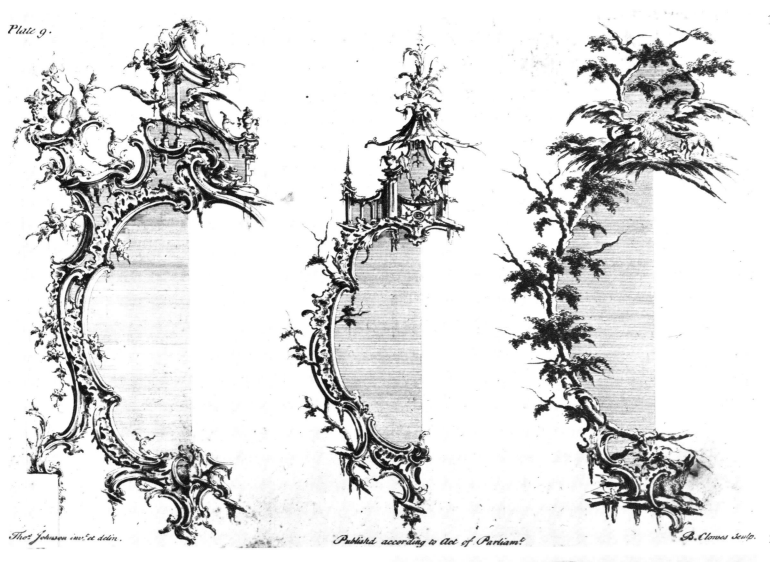

Plate 9.

Thos. Johnson inv.t et delin.   Publish'd according to Act of Parliam.t   B. Clowes sculp.   **301**

301. Drawings by Thomas Johnson, English, circa 1760. These drawings for oval mirrors depict imaginery pastoral landscape scenes bent round to accommodate a panel of glass. The drama of nature is captured in the hunt theme to the right and left. In contrast, the center drawing is a peaceful interpretation of a serene Chinese landscape painting. Thomas Johnson was one of the originators of the exotic Rococo style as he published such drawings in 1758 and 1761. He was also a master carver whose work was of fine quality (see also #244.)

302. Gilded wood, English, circa 1770. The pastoral theme is obvious in this mirror with the phoenix poised over the innocent sheep. The design and carving are quite open and lacy, using the accustomed Rococo motifs of c-scrolls, cabochon, icicles, ruffles and foliage to create the completed design.

302

303

304

305

306

307

308

303 to 310. Gilded wood, English, circa 1760 to 1770.

309

310

311

312

313

311. Painted and gilded wood paneling and glass mirrored doors, English, circa 1790. The Gothic Rococo style was a late development of this period. This room was designed by architect James Wyatt for Thomas Barret's home in England. The best known house in this Gothic Rococo style was Horace Walpale's "Strawberry Hill"; therefore, the term "Strawberry Hill Gothic" is sometimes used.
312 and 313. Gilded wood, English, circa 1780. This Gothic Rococo mirror and matching pair of sconces (one shown) illustrate a typical interpretation of this style.

Fretwork Gilded

314. Mahogany veneer and gilded wood, English, circa 1750.
This next group of thirty-two mirrors occupy a transitional position in the evolution of design between the architectural mirrors and the fret-carved group. Unlike the architectural mirrors, they do not have complete outlines in gold in the lower crest and they do not have architectural corners in the upper crest. They are, however, usually veneered with walnut or mahogany, and have shaped outlines like fret-work mirrors with gilded caps on the upper corners of the frame.

315. Mahogany veneer and gilded wood, English, circa 1750. The gilded foliage above the top crest is supplemented by the phoenix and unusual applied decorations in both top and bottom crests. Also unusual is the piercing in the side ears of the fretwork frame. *Courtesy of David Stockwell, Inc.*

316. Walnut veneer and gilded wood, English or Northern European, circa 1760, 49" high, 27" wide.
We look at this mirror as an experimental model combining architectural pediment and swags, central cartouche, and very unusual pierced, carved and gilded inserts in the lower and upper crests. These elements are not particularly compatible and the design was not repeated. *Courtesy of Alfred Bullard, Inc.*

314

315

316

317

319

320

318

317 and 318. Mahogany veneer with gilded wood, English or American, possibly by John Elliott of Philadelphia, circa 1770, 43½" high, 20½" wide. The rather primitive gilded pediment caps draw immediate attention to this mirror which otherwise is a good representative of the English fret-work group.

The construction of the back of the frame has side braces flanking the back board running straight up into the crest. Many English mirrors have side braces cut at a 45° angle at these corners to join the horizontal brace at diagonal junctions. Here, instead, the horizontal brace is cut straight to butt against the extending side braces. The label of John Elliott of Philadelphia has been found on mirrors of this construction (as examples in the fret-work section of this book present). Whether he imported them from England or made any or all in Philadelphia is not certain. *Courtesy of Herbert Schiffer Antiques.*

319. Mahogany veneer and gilded wood, English, circa 1750, 45" high, 22¾" wide. The architectural crest has an unusual gilded edge on the curve between the scrolls and the basket, presumably original. This detail may be unique here. *Courtesy of Herbert Schiffer Antiques, Inc.*

320. Mahogany veneer and gilded wood, English, circa 1760. The gilded caps are a very successful addition to this fret-work mirror. *Courtesy of Herbert Schiffer Antiques.*

321, 322 and 323. Mahogany veneer with gilded wood and incised gilt decoration, English, circa 1760. These three mirrors are variations of the fretwork style, each with gilded caps and very unusual and fancy incised floral decoration in the base and crest. The carving and gilt are of fine quality indicating that these mirrors probably were made at the same time and by the same carvers as the fancier all-gilt frames. *Courtesy of Herbert Schiffer Antiques.*

321   **322**   **323**

324

325

324. Mahogany veneer and gilded wood, English, circa 1760, 52" high, 20½" wide. The extra ornament of the incised carving gives this mirror additional elegance. *Courtesy of David Stockwell, Inc.*

325. Mahogany veneer and gilded wood, English, circa 1770. *Courtesy of Herbert Schiffer Antiques, Inc.*

326. Mahogany veneer and gilded wood, English, circa 1770, 48" high, 26" wide. *Courtesy of the Diplomatic Receptioms. United States Department of State.*

327. Mahogany veneer and gilded wood, English, circa 1770. *Courtesy of Alfred Bullard, Inc.*

328 and 329. Mahogany veneer and gilded wood, English, circa 1770. *Courtesy of Herbert Schiffer Antiques.*

330. Mahogany veneer and gilded wood, English, circa 1760, 47" high, 24" wide. *Courtesy of Herbert Schiffer Antiques.*

326

327

328

329

330

Round Corners

The following group of mirrors derived from the Baroque veneered style associated with the Queen Anne period in England and America. (see #87–100 in this book.) The frames are solid or veneered mahogany or walnut with shaped outline usually including scrolled "ears" at the top and bottom corners. This type of frame is known as fret-work. During the Rococo period in the mid-eighteenth century, fret-work mirror frames were decorated with a gilded and carved gesso band between the glass and frame, known as a fillet. Also, side garlands of leaf and/or fruit swags and gilded carving in the crests were often included.

The group can be divided into two sections by the design of the upper corners of the inside edge of the frame at the mirrored glass panels. The earlier (generally pre-1770) section has rounded corners in the frame around the glass and the later (generally post-1770) has squared corners in the frame around the glass. The lower corners inside the frame can be rounded or square in either group.

331

332

333

334

331. Mahogany veneer on white cedar and white pine, American Delaware Valley, circa 1730, 67" high.

This is a dynamic mirror to introduce this group because it is an American mirror with lovely proportions and excellent quality carving and gilding. The shape is certainly derived from the Baroque (Queen Anne) tradition with arched crest, but here the crest is fret cut with a pierced and gilded foliate medallion. The secondary woods (cedar and white pine) conclusively identify its origin as American since these woods do not grow in Europe but are prevalent in Eastern America (particularly in the valley of the Delaware River near Philadelphia) and were convenient for cabinet makers to obtain. They are the most frequently used secondary woods on all early American furniture. To complicate the issue, mirrors of English manufacture have been found to have white pine secondary wood. Since it is a soft and easily worked wood, it was imported to England from America in the eighteenth century. No one fact can determine the origin of a piece of furniture. *Courtesy Craig and Tarlton, Raleigh, North Carolina.*

332. Walnut veneer, English, circa 1730-1760, 35" high, 20¼" wide.

The double beveled glass with glass star attached is a very unusual survivor of Queen Anne glass framed mirrors. Here the frame is a simple fret-work design with pierced crest. *Courtesy of Tryon Palace Restoration, New Bern, North Carolina.*

333. Walnut veneer, English, circa 1750, 35½" high, 20 5/8" wide.

The outline of the glass panel is made more interesting by the scalloping of the fillet. *Courtesy of Herbert Schiffer Antiques.*

334. Mahogany veneer and gilded wood, English, circa 1750, 43½" high, 25" wide. *Courtesy of Herbert Schiffer Antiques.*

335. Mahogany veneer and gilded wood, American or English, circa 1730-1750, 47" high, 26" wide. The gilded ears on this mirror are an unusual variation giving rise to its design being made in the colonies of England rather than in the mother country. However, the carving is superior and would be expected only from a London craftsman's shop. *Courtesy Israel Sack, Inc., New York City.*

336. Mahogany with gilded wood, American, Philadelphia, circa 1760. Many bureau book cases or secretaries were built with mirrored doors in England and America during the eighteenth century. This splendid example is representative of the form which extended through each of the furniture styles.

335

336

337

338

339

337. Mahogany veneer and gilded wood, English, circa 1760. The gilded intaglio carving adds a Rococo spirit to this well proportioned mirror. *Courtesy of Alfred Bullard, Inc.*

338. Mahogany veneer and gilded wood, English, circa 1750, 29 1/8" high, 17" wide. *Courtesy of Herbert Schiffer Antiques.*

339. Mahogany veneer and gilded wood, English, circa 1750. *Courtesy of Herbert Schiffer Antiques.*

340

342

341

340 and 341. Mahogany and white pine, American, bearing the label of Joseph White of Wilmington, Delaware, circa 1785, 51 3/8" high, 26½" wide. This mirror may have been made in Philadelphia or New York, although it bears the label of Joseph White of Wilmington, Delaware who probably sold it. White was a druggist and mirror merchant. (Another label is shown #379.) *Courtesy, The Henry Francis du Pont Winterthur Museum.*

342. Walnut veneer, inlay, and gilded wood, English, circa 1760. The intaglio carved and gilded leaves in the crest add fanciness to this fret work mirror, but the line of light and dark contrasting wood inlay is the real surprise. Inlay is generally associated the the Classical style at the end of the eighteenth century. Its use here in conjunction with gilded carving is unexpected. *Courtesy of Alfred Bullard, Inc.*

343

345

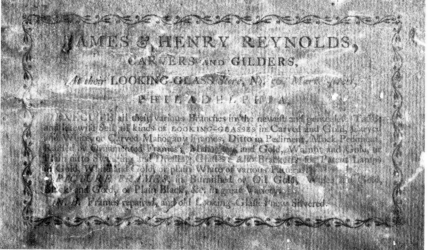

344

343 and 344. Mahogany and gilded wood, American, bearing the label of James and Henry Reynolds, circa 1795, 49¼" high, 25 1/8" wide. James Reynolds (1736-1794) was listed in Philadelphia directories at various addresses between 1785 and 1794. He imported English, French and Dutch mirrors, but also is credited with making the mirror shown as #246 in this book. His sons James and Henry continued his business. This mirror has an elegant, tall proportion and refined carving. *Courtesy, The Henry Francis du Pont Winterthur Museum.*

345. Mahogany and gilded wood, English, circa 1760. The combination of the gilded eagle in the top crest and foliage in the lower crest is an unusual and successful design. *Courtesy of Alfred Bullard, Inc.*

346. Walnut veneer and gilded wood, English, circa 1760. *Courtesy of Herbert Schiffer Antiques.*

347 and 348. Mahogany on pine and gilded wood, English, circa 1760, bearing label of William Bittle and James Cooper, Boston. This English mirror was made about 1760 and sold by Bittle and Cooper about 1825 in Boston. One cannot jump to conclusions that either a label is original or contemporary with a mirror. Knowledge of styles and sellers or makers must combine logically in order to make attributions based on labels. *Courtesy of Taylor and Dull, New York.*

349. Mahogany veneer and gilded wood, English, circa 1760. The variations of eagles and phoenixes is an interesting study; this one is well proportioned for the crest size. *Courtesy of Herbert Schiffer Antiques.*

346

347

349

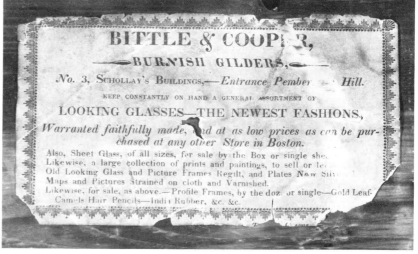

BITTLE & COOPER,
—BURNISH GILDERS,—

No. 3, Schollay's Buildings,—Entrance Pember___ Hill.
KEEP CONSTANTLY ON HAND A GENERAL ASSORTMENT OF

LOOKING GLASSES—THE NEWEST FASHIONS,

Warranted faithfully made, and at as low prices as can be purchased at any other Store in Boston.

Also, Sheet Glass, of all sizes, for sale by the Box or single she___
Likewise, a large collection of prints and paintings, to sell or le___
Old Looking Glass and Picture Frames Regilt, and Plates New Silv___
Maps and Pictures Strained on cloth and Varnished.
Likewise, for sale, as above.—Profile Frames, by the doz___ or single—Gold Leaf
Camels Hair Pencils—India Rubber, &c. &c.

348

350

351

352

353

354

355

356

350. Mahogany veneer on pine, English, circa 1760, 36¼" high, 24½" wide. The cut outs in this design create the entire decoration, there is no gilding. A similar cut-out fret work mirror with gilded caps is shown in # *Courtesy of Herbert Schiffer Antiques.*

351. Mahogany veneer and gilded wood, English, circa 1760. The cut out in the lower crest matches the cut out in #350, but here gilded sections are also included. *Courtesy of Herbert Schiffer Antiques.*

352. Mahogany veneer on pine and gilded wood, English, circa 1760. *Courtesy of Herbert Schiffer Antiques.*

353. Mahogany veneer on spruce or Scotch pine, English, circa 1762-1763 bearing part of the 3rd or 4th label of John Elliott, Senior, of Philadelphia. John Elliott was an English cabinetmaker who came to America by 1758 when he imported looking glasses from England and sold them in Philadelphia. Ten different labels have been identified from his shop during the late eighteenth century when father John, Senior and son John, Junior ran the business. The first four labels were printed in German and English texts and were used over the period from 1758 until about 1784. Then son John, Junior ran the business and made mirrors and these labels bear his name. The last six labels are printed in English only and were used between 1784 and 1810.

In this book, when an Elliott label appears on a mirror, the label is shown and the dates of the label's use are given. (see Mary Ellen Hayward's Masters' thesis *The Elliotts of Philadelphia,* University of Delaware program at Winterthur, 1971.)

354. Mahogany and gilded wood, English, circa 1760, 44" high, 22" wide. *Courtesy of Herbert Schiffer Antiques.*

355. Mahogany and gilded wood, English, circa 1760. *Courtesy of Herbert Schiffer Antiques.*

356. Mahogany veneer on oak and gilded wood, English, circa 1760. *Courtesy of Herbert Schiffer Antiques.*

358

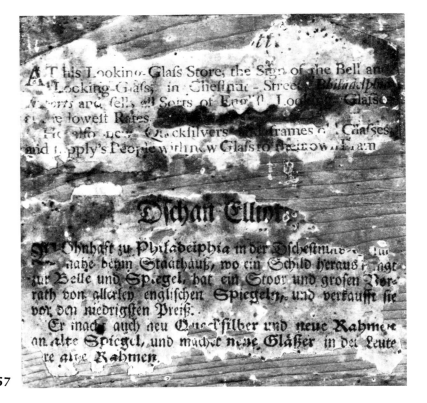

357

357 and 358. Mahogany veneer on spruce or Scotch pine, and gilded wood, English, bearing the first label of John Elliott, Senior, of Philadelphia who sold it, circa 1758, 32½" high, 15" wide. This type of Queen Anne style mirror is known to have been made well into the middle of the eighteenth century and Rococo style. It just did not go out of fashion. This particular example bears the first label of the merchant in Philadelphia who imported and sold the mirror, John Elliott, Senior (See also #353 and other mirrors with his labels in this book.) The remnant of the label (#358) shows both German and English text. This is identified as the first label by the location. "Chestnut Street" for Elliott's shop, and Elliott's first name printed "Dichan" in the German. *Courtesy, The Henry Francis du Pont Winterthur Museum.*
359 and 360. Mahogany veneer on spruce, English, bearing the second label of John Elliott, Senior of Philadelphia who sold it, circa 1762. This is another Queen Anne style mirror made late in the eighteenth century in England and imported to Philadelphia by John Elliott, Senior. This label (#360) is similar to the first shown (#358) but Chestnut Street has been scratched out of both the German and English texts and the word "Walnut" written in. John Elliott, Senior moved his shop from Chestnut Street to Walnut Street in 1762. While his supply of labels lasted, he simply wrote over the old address.

359

360

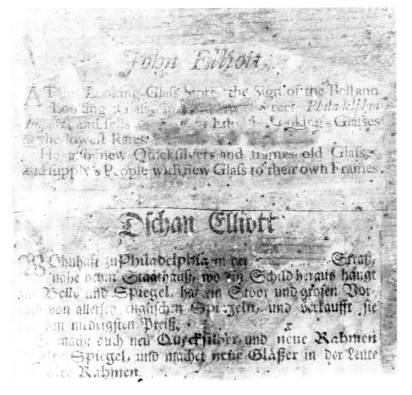

361

363

362

361 and 362. Mahogany, English, bearing the third label of John Elliott, Senior of Philadelphia who sold it, circa 1763, 21½" high, 10½" wide. The mirror is a simple Queen Anne style imported to this country well into the middle of the century. It bears the third label of John Elliott, Senior. His address is now "Wallnut" (with two l's) Street, and his name in German is spelled "Johannes." This label was used about 1762-1763. *Courtesy of the Anglo-American Art Museum, Louisiana State University.*

363 and 364. Mahogany veneer on spruce or Scotch pine, English, bearing the fourth label of John Elliott, Senior, who sold it, circa 1763. The label is the fourth of John Elliott, Senior with "Walnut" spelled with one "l" and the German word "Waaren lag" Store. This label was used about 1763 until 1784 when the Walnut Street location was sold. Up until this time, the mirrors were mostly imported, as their woods are of European origin. Elliott may have made some of them in America before 1777, but in that year American non-Importation agreements precluded further foreign mirrors entering America. Elliott made his stock from then onward. *Courtesy of David Stockwell, Inc.*

364

365

367

366

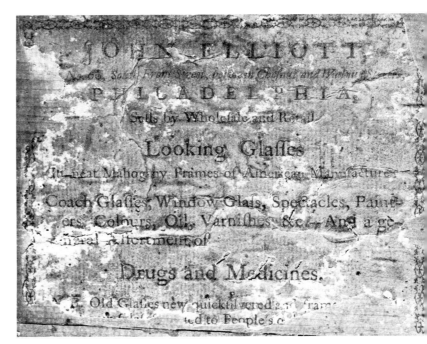

368

365 and 366. Mahogany veneer on white pine, American, bearing the third label of John Elliott, Junior who made it, circa 1796-1804.

John Elliott, Junior continued his father's business after 1784 and used a new label, all in English, with his name at the top. His first label used the address "on the West side of Front-Street, between Chestnut and Walnut Streets." His second label uses the address "No. 60, South Front Street, between Chestnut and Walnut Streets."

#366 shows his third label which resembles the second label in the address, but "Looking Glasses in neat mahogany frames of American Manufacture" has been added and the border decoration is a spiral design.

His fourth label is identical with this third, but the border decoration is a dot design. *Courtesy of Herbert Schiffer Antiques.*

367 and 368. Mahogany veneer, American, bearing the fifth label of John Elliott, Junior, who made it, circa 1796-1804, 39 9/16" high, 21" wide. The mirror has rounded top corners of the frame adjacent to the glass, an unusual feature for mirrors of this date which were mostly squared here. The label matches John Elliott, Junior's third and fourth labels in the lettering, but on this fifth label the border decoration is a geometric pattern. *Courtesy of The Philadelphia Museum of Art. Photograph by A.J. Wyatt, staff photographer.*

369 and 370. Mahogany and white pine, American, bearing the label of John Elliott and Son, circa 1804-1810. This looking glass resembles ones labeled by James Stokes in Philadelphia at this same time. It is not known whether Stokes made mirrors sold in Elliott's Shop, or if a third party made mirrors sold by both Elliott and Stokes. The label is the last one used by the Elliott firm. It reads "John Elliott and Son" and lists drugs and medicines, window and picture glass and paints among their stock as well as the looking glasses. *Courtesy of David Stockwell, Inc.*

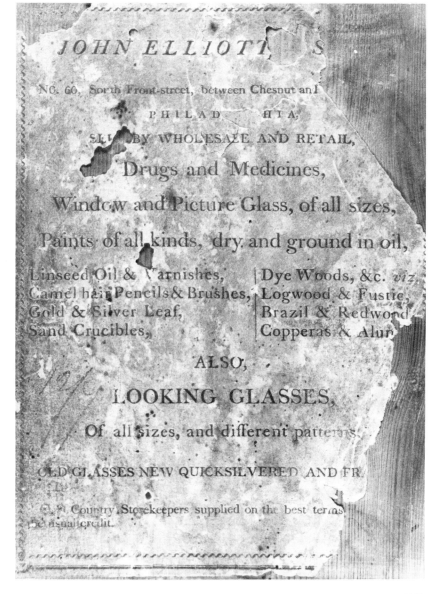

369

370

371

372

373

374

375

371. Mahogany veneer on spruce or Scotch pine, English, bearing a label [too fragmented to identify further] of John Elliott, Senior, who sold it in Philadelphia, circa 1762. This style is now associated with the Elliott firm as so many labeled examples have the same low crest and fret-work cut outs. *Courtesy of David Stockwell, Inc.*
372. Mahogany veneer on spruce or Scotch pine, English, bearing the fourth label of John Elliott, Senior, who sold it, circa 1763. *Courtesy of David Stockwell, Inc.*
373. Both are mahogany on spruce or Scotch pine, English, each bearing labels of John Elliott, Senior, who sold it, circa 1763. These are two of the typical mirrors which bear Elliott labels. They represent the Queen Anne type with no lower crest, and more usual fret work type of Rococo design. *Courtesy of David Stockwell, Inc.*

374. Mahogany veneer on spruce or Scotch pine, English, bearing the third label of John Elliott, Senior, who sold it, circa 7162, 19¼" high, 10½" wide. *Courtesy of David Stockwell, Inc.*
375. Mahogany veneer, English, bearing the third label of John Elliott, Senior, who sold it in Philadelphia. *Courtesy of Herbert Schiffer Antiques.*
376. Walnut veneer on pine, English, circa 1760. *Courtesy of Herbert Schiffer Antiques, Inc.*

376

377

378

379

377. Walnut veneer, English, circa 1750, 32¼" high, 18¼" wide. Walnut veneer was used before mahogany veneer, but continued simultaneously into the last quarter of the eighteenth century. *Courtesy of Herbert Schiffer Antiques.*

378 and 379. Mahogany veneer on white cedar and tulip poplar, American, bearing a label of Joseph White who sold it, circa 1760-1780, 49 1/8" high, 27¼" wide. This fret work mirror is very similar to labeled Elliott mirrors, and may have a common source. White died in 1798. The label (#379) reads "Sold at/Joseph White's Apothecary Shop,/At the Boy and Mortar,/Market Street,/Wilmington." This label probably pre-dates the Joseph White label shown as #341 in this book. *Courtesy, The Henry Frances du Pont Winterthur Museum.*

380

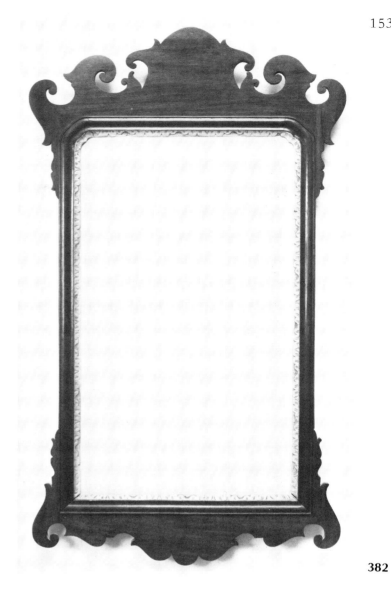

382

381

380 and 381. Mahogany veneer on spruce, English, bearing the label of Thomas Aldersey, London, circa 1760. In style this is very similar to mirrors sold by the Elliotts in Philadelphia at this period, and perhaps Thomas Aldersey made some of those mirrors. It is very unusual to find a paper label on an English mirror; this one .(#381) reads in part, "Thomas Aldersey,/Looking Glass Maker,/(torn). *Courtesy of David Stockwell Antiques.*

382. Mahogany and gilded wood, English, bearing the fourth label of John Elliott, Senior, who sold it in Philadelphia, circa 1763-1777, 38½" high, 21 1/8" wide. *Courtesy, The Henry Francis du Pont Winterthur Museum.*

154

383

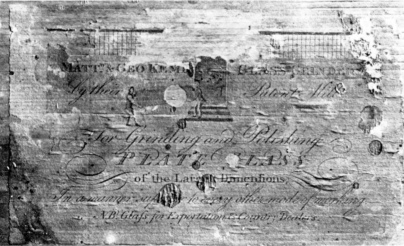

384

383 and 384. Mahogany veneer and gilded gesso, English, bearing the label of Matthew and George Kemp, circa 1790. George Kemp was a cabinetmaker and glassgrinder in London from about 1760 to 1790. Matthew was apparently his son. In 1790, the Plate Glass Company at Ravenhead "agreed to provide for Messieurs Kemp all Materials and Things needful for building a Mill or, Mills, according to their said Secret or Invention' for polishing and grinding plate glass" (see Wills. *English Looking Glasses*, page 154.) This label reads "Mattw & Geo. Kemp glass grinders/by their Patent Mill/for grinding and polishing/Plate Glass/of the largest

385

Dimensions/in a manner superior to every other mode of working/N.B. glass for Exportation & Country Dealers."

Therefore, was the glass for this mirror ground at the Kemp mill? Is that what the label is trying to tell us? The frame maker is apparently unknown. His work is very similar to the other mirrors in this section, labeled or not.

385. Mahogany, American, bearing the fourth label of John Elliott, Junior, circa 1796-1804. *Courtesy, The Henry Francis du Pont Winterthur Museum.*

386. Mahogany, English or American, circa 1770. *Courtesy of Herbert Schiffer Antiques, Inc.*

387. Mahogany veneer on white pine, American, circa 1790, 41 1/8" high, 22" wide. So many mirrors of this fret-work design and combination of mahogany and white pine have been found in and near Philadelphia and with Philadelphia labels (Elliott, Stokes), that an attribution here to the Philadelphia area seems logical. *Courtesy of Herbert Schiffer Antiques.*

388 and 389. Mahogany veneer, English or American, circa 1770. *Courtesy of Herbert Schiffer Antiques.*

386 **387**

388 **389**

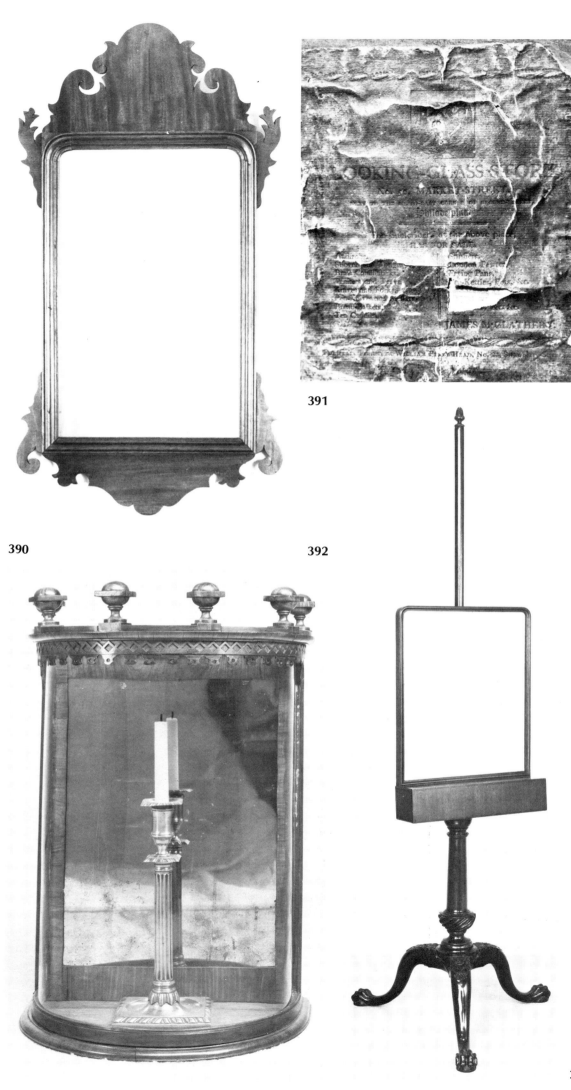

391

390

392

390 and 391. Mahogany veneer on pine, American, bearing the label of James McGlathery, Philadelphia, circa 1805. This mirror's frame outline matches those of mirrors labeled by James Stokes in Philadelphia and his successors Wayne and Biddle. Only the rounded upper corners at the inside frame are present on just this mirror (the others having square corners here.) This label reads "Looking Glass Store/No. 50 Market Street/next to the southeast corner of Second Street/Phila-delphia./The Subscriber at the above place/has for sale/(list) Andirons. Shovels and Tongs. Brass Candlesticks. Waiters and Trays. Knives and Forks. Pan-knives and Scissors. Bread Bas-kets. Tea Caddies, Snuffers, London Pewter, Frying Pans. Tea Kettles, Pots. Ec. (torn) Ec., Ec./James McGlathery/Folwell Printers, at William Penn's Head, No. 23. Strawberry Street."

The printer, Richard Folwell, was in business in Philadelphia between 1794 and 1809, but had his shop on Strawberry Street only in 1805. This fact, then, probably dates the mirror very nearly to that year.

However, nothing is said about a mirror maker. Did some anonymous third party make this and the similar Stokes mirrors? Or, did Stokes make this and sell it through Mc-Glathery—or vice-versa?

Because so often fret-work mirror frames have nearly matching crest cut-outs, it can be suggested that templates were made and used over and over again, perhaps by more than one maker and with minor variations as frequent innova-tions. The Elliott and Stokes mirrors particularly suggest this template theory as a possibility. *Courtesy of Herbert Schiffer Antiques.*

392. Wall lantern with mahog-any frame and mirrored back, English, circa 1760. Survivors of this style lantern are rare today, yet many must have been made. The basic useful shape is ornamented only by the blind fretwork and unusual finials which actually give the lantern an appealing form. *Courtesy of David Stockwell, Inc.*

393. Mahogany pole screen or dressing stand with mirrored panel on carved base, English, circa 1760, 60" high. Although this is a very rare piece today, its usefulness can be imagined. Others must have been made but have not survived. *Courtesy of the Victoria and Albert Museum.*

393

Square Corners

394. Mahogany, American, associated with James Stokes, Philadelphia, circa 1790. The high arching crest and tightly scrolled curls are established as trademarks of mirrors associated with James Stokes' label (see #412). Stokes appears only to have sold the mirrors, the makers remain unknown.

This mirror has amateur scratch carving in the crest reading "W.P. 1772" and a long history in Chester County, Pennsylvania where William Pusey was a prominent resident at this time. *Courtesy, Mrs. Herbert Schiffer.*

395. Mahogany and light wood inlay, American, possibly by John Elliott, Junior, dated 1786. The crest is very similar to crests on labelled Elliott mirrors, and at this time John, Junior was making the mirrors he sold. The inlay in the crest is beautifully designed and executed, bearing the date 1786 and initials M.T. The needlework sampler is signed by Mary Thomas who lived in Chester County, Pennsylvania at this time. No doubt this frame was made for this sampler by a local mirror frame maker. *Courtesy of Mr. and Mrs. Frances Edward Judson.*

396. Mahogany, American, Pennsylvania, circa 1796. It is difficult now to speculate whether the needlework was made for the frame or vice-versa. Several of these old frames remain with needlework so beautifully matched that one feels they must have been together originally. This sampler is signed "Ann Wickersham, her sampler made in the fifteenth year of her age A.D. 1796" The frame would have been made by a local craftsman—such as one who made mirrors. (See #398 inscription about framing needlework.) *Courtesy of Mrs. Herbert Schiffer.*

394

395

396

397

399

398

397 and 398. Mahogany and white pine, American, bearing the label of Bernard Cermenati, Newburyport, Rhode Island, circa 1807 to 1809, 17 5/8″ high, 10 3/8″ wide. The outline of the solid mahogany crest will be seen on other examples of Cermenati's work. This mirror has a provincial Queen Anne style, yet was made at the beginning of the nineteenth century. This label was used by Bernard Cermenati for only the short time he was in Newburyport, after which he worked in Salem, Massachusetts (1810), Boston (1811), Portsmouth, New Hampshire (1812), and Boston again (1813-1818 when he died). Notice is drawn to the last few lines of his label (#398) which reads "Ladies Needlework handsomely framed ia (sic) the most/modern style, and the shortest notice." *Courtesy, The Henry Francis du Pont Winterthur Museum.*

400

401

402

403

399. Mahogany veneer and pine, American, bearing the latest label of John Elliott and Son, Philadelphia, circa 1804 to 1810, 38¾" high, 19¾" wide. *Courtesy of Herbert Schiffer Antiques.*
400. Mahogany and pine, American, bearing the label of John Elliott, Junior of Philadelphia, circa 1790. *Courtesy of Herbert Schiffer Antiques.*

401. Mahogany veneer on white pine and tulip poplar with gilded wood, American, bearing the label of Kneeland and Adams, Hartford, Connecticut, circa 1792 to 1795. *Courtesy, Museum of Early Southern Decorative Arts, winston-Salem, North Carolina.*
402 and 403. Mahogany veneer on white pine and tulip poplar with gilded wood, American, bearing the label of Kneeland and Adams, Hartford, Connecticut, circa 1792 to 1795, 37 5/8" high, 18¾" wide. Mirrors #401 and 402 were both made by the firm Kneeland and Adams of Hartford, Connecticut, and although they have their differences, they both have "fish tail" cut outs at the ears in all four corners. The fret cut outline of both gives a lacy overall effect.

Samuel Kneeland advertised a variety of furniture and framed gilded and silvered looking glasses in 1786 [*American Mercury*]. He and Lemuel Adams were partners between September, 1792 and March 1795 after which they each worked alone in Hartford. The label, #403, shows quite fancy furniture including a looking glass. These pieces may have been assembled by the label printer Elisha Babcock and have little resemblance to the work of Kneeland and Adams, as has been the case of labels of other cabinetmakers (see E. & R. Dubrow, *American Furniture of the 19th Century, 1840-1880*, pages 69 to 71.) *Courtesy, The Henry Francis du Pont Winterthur Museum.*

404

405

404. Mahogany, American, bearing label of Nicholas Geffroy, Newport, Rhode Island, circa 1800, 25" high, 12¾" wide. The contrasting light stringing inlay in the frame of this mirror is the first instance so far of this decoration which was to become standard in the following Sheraton or Federal style. The maker, Nicholas Geffroy (1761-1839) worked variously as a watch maker, silversmith, joiner, looking glass maker, and merchant. A watchpaper label of his lists him as a watch maker at Thames Street, and lists Seals, Keys and Trinkets. An 1809 newspaper advertisment lists a wide variety of items for sale. *Courtesy of Los Angeles County Museum of Art. Colonel and Mrs. George J. Deuis Fund.*

405 and 406. Mahogany and pine, American, bearing the label of Jos. & Jno. Vecchio, New York, circa 1805. This mirror is just a slightly more elaborate version of #404 by Nicholas Geffroy, and its similarity hints at the existence of pattern books and/or templates for the fashionable mirrors. The makers, Joseph and John Del Vecchio (English version of their label names, see # 406,) arrived in New York with their brothers Francis and Charles from Moltrasio, Italy in 1800. Joseph and John had a store, looking-glass, and frame shop in Albany, New York in 1803-1804. In 1805 they moved to 136 Broadway, New York City where they had a print and looking glass store (label 406) until 1810. They moved to 134 Broadway in 1812, and each died separately in 1815. *Courtesy of M. Finkel and Daughter.*

406

407

409

408

407 and 408. Mahogany veneer and white pine, American, bearing the label of John Townsend, Newport, Rhode Island, circa 1790, 41 7/8" high, 21½" wide. The pine-cone shape of the top ears is a detail usually associated with mirrors by James Stokes of Philadelphia. The flat plinth at the crest was probably made to hold a carving or gilded element. John Townsend (1732-1809) was a cabinetmaker in Newport, Rhode Island between 1764 and 1809 and his wife named Philadelphia was the daughter of artist Robert Feke. *Courtesy, The Henry Francis du Pont Winterthur Museum.*

409. Mahogany, American, New England, circa 1800, 44" high, 25" wide. The oval eagle inlay in the crest is usually associated with pieces of New England origin, and this example has sixteen stars arching at the top. The contrasting light and dark stringing inlay became typical of this period. See also mirror 442. *Courtesy, The Diplomatic Reception Rooms, United States Department of State.*

410

412

411

413

410. Mahogany, American, circa 1800, 29½" high, 15¾" wide. The shell patterned oval inlay is a Classical element and together with the square corners and stringing outline help to determine the date of this mirror after 1790. *Courtesy of David Stockwell, Inc.*

411 and 412. Mahogany veneer on pine, American, bearing the label of James Stokes of Philadelphia who sold it, circa 1791 to 1804, 28 7/8" high, 16 7/16" wide. A typical feature of mirrors bearing Stokes labels are the tightly scrolled jigsaw cut-outs in the frame. The lines of inlay on this example are not typical. Small mirrors bearing the Stokes label tend to be made of solid mahogany, not veneer. The label (412) bears his address between 1791 and 1804, and as it states, James Stokes was a merchant who probably never made anything he sold. He had shops at various addresses in Philadelphia between 1785 and 1810 when he was succeeded by his two sons-in-law, Caleb Parry Wayne and Charles Biddle, Jr. (Wayne and Biddle). *Courtesy of Frank Schwarz & Son.*

413. Mahogany, American, associated with James Stokes of Philadelphia, circa 1795. Besides the tightly scrolled volutes and high arching crest, the pine-cone shaped ears are usually found on mirrors bearing James Stokes labels. *Courtesy of Herbert Schiffer Antiques, Inc.*

414

415

416

414. Mahogany and white pine, American, bearing the label of James Stokes of Philadelphia who sold it, circa 1795, 28 7/8" high, 16 7/16" wide.
415. Mahogany veneer, American, associated with James Stokes of Philadelphia who sold it, circa 1795. *Courtesy of Herbert Schiffer Antiques, Inc.*
416. Mahogany and pine, American, bearing the label of Wayne and Biddle of Philadelphia who sold it, circa 1815. Caleb Wayne and Charles Biddle succeeded their father-in-law James Stokes as merchants in 1811. The form of their mirrors continued the details associated with Stokes. *Courtesy of Herbert Schiffer Antiques, Inc.*

417

419

418

420

417 and 418. Mahogany and pine, American, bearing the label of Wayne and Biddle of Philadelphia who sold it, circa 1815. Wayne and Biddle (see 416) ran the store between 1811 and 1822 after which Caleb Wayne continued in business at various Philadelphia locations until 1833. Charles Biddle lived in Philadelphia until 1826, and the following year was admitted to the bar in Nashville, Tennessee.

This label (418) includes the line "Successors to James Stokes". *Courtesy, The Henry Francis du Pont Winterthur Museum.*

419 and 420. Mahogany, American, bearing the label of Wayne and Biddle of Philadelphia who sold it, circa 1820, 21" high, 12 5/8" wide. This label varies from *418,* being a later version without "Successors to James Stokes" and with a listing of their merchandise as a border (see also *586, 587,* and *588). Courtesy, The Henry Francis du Pont Winterthur Museum.*

421

423

422

424

421 and 422. Mahogany with gilded gesso, American, bearing the label of Cermenati & Monfrino, Boston, 1806, 36¼" high, 18¼" wide. This mirror introduces a group made in Boston and north of Boston between 1806 and 1809 by three firms involving Paul Cermenati. He used the same patterns in each of his partnerships. The mirrors are alike in their solid mahogany construction, and flat carved and dull gilded pressed composition eagle or other crest decoration.

In 1806, Paul Cermenati and G. Monfrino advertised as carvers and gilders at 2 State Street, Boston. Late that year, Cermenati was advertising a different partnership, this time with John Bernarda. In 1807 they moved to Salem, Massachusetts. By 1809, Paul Cermenati was working in Boston under the name P. Cermenati and Co.

423 and 424. Mahogany with gilded gesso, American, bearing the label of Cermenati and Bernarda of Boston, 1806-1807, 25 1/8" high, 13 1/8" wide. The gilded feathers appear on these mirrors frequently with the same fret-work outline as those with gilded eagle decoration. The label (424) lists a wide variety of pictures from Europe and frames made by themselves. *Courtesy, The Henry Francis du Pont Winterthur Museum.*

425

427

425. Mahogany and gilded gesso, American, attributed to Paul Cermenati, Boston, circa 1806, 35″ high, 17¼″ wide. *Courtesy of Herbert Schiffer Antiques, Inc.*
426. Mahogany and gilded gesso, American, Boston area, circa 1810, 26¼″ high, 14″ wide. *Courtesy of Herbert Schiffer Antiques, Inc.*
427. Mahogany and gilded gesso, American, Boston area, circa 1810. *Courtesy of Herbert Schiffer Antiques, Inc.*

428 and 429. Mahogany and gilded gesso, American, bearing the label of Edward Lothrop of Boston, circa 1820, 27¼″ high, 14 5/8″ wide. This mirror looks very similar to the Cermenati group, and perhaps there is more than a casual connection. Edward Lothrop is listed as a gilder and looking glass and frame maker in Boston from 1813 through 1836. This label (429) has the 28 Court Street address which he used in advertisements from 1820 until 1823. (See also 556) *Courtesy of Herbert Schiffer Antiques, Inc.*

430. Mahogany, American, Boston area, circa 1810, 21½″ high, 12½″ wide. *Courtesy of Herbert Schiffer Antiques, Inc.*

426

428

430

KING-GLASSES CLEAN
DWARD LOTHROP,
his Old Stand No. 25 Court Street
BOSTON
..Near Concert Hall..

TINUES to manufacture Elegant Burnished
ILT LOOKING-GLASSES,
d in the best manner, under his immediate in-
, which will be found (we need not say) vastly
rior to those that are urged upon the public every
at Auction, and by others not acquainted with the
de. ALSO, On Hand at all Times,
A good Supply of low-priced Gilt & Mahogany framed
GLASSES
Well suited to the Country Market; which will be sold
extremely Cheap.
Traders from the country, who buy to sell again, have
found, and, we trust, will still find it for their interest
to call as above, where they will find good Stock, and
carefully packed up for transportation in the best manner.
N.B. Looking-Glasses new Framed, together with
Portrait Frames, Needlework, and all sorts of Pictures,
for a reasonable compensation.

429

431

433

ELISHA TUCKER,
CABINET AND CHAIR
MANUFACTURER,
RESPECTFULLY informs his Friends and the Publick, that
he Manufactures and offers for Sale on reasonable terms, at No.
40, MIDDLE-STREET...BOSTON, a general assortment of
CABINET FURNITURE and CHAIRS.
Mahogany Looking-Glass Frames
of all sizes, executed in the neatest manner and at the shortest notice.
N. B. No exertions shall be spared which will serve to render
satisfaction to those who may please to favor him with their
friends.

432

434

431 and 432. Mahogany, American, bearing the label of Elisha Tucker of Boston, circa 1810, 17 9/16" high, 11½" wide. This mirror is closely related to the Edward Lothrop (428) and Cermenati type (421 to 427). Elisha Tucker (1784-1827) made furniture, chairs, and frames at 40 Middle Street, Boston in 1809 and 1810 and at various other addresses in Boston until 1827. *Courtesy, The Henry Francis du Pont Winterthur Museum.*

433 and 434. Mahogany with satinwood, tulip poplar, and ash, American, bearing the label of Wells M. Gaylord of Utica, New York, circa 1828, 37 1/8" high, 22 1/8" wide. This is quite an unusual mirror. The fret work frame surrounds a plainer, rectangular frame which became very popular in the 1830's. The address on the label, No. 55 Genesee Street is recorded as his address in 1828. He moved to several other locations in Utica until 1845 when his business seems to have been taken over by Edwin Gaylord. *Courtesy, The Henry Francis du Pont Winterthur Museum.*

Transitional

435. Mahogany veneer and gilded wood, English or American, circa 1790, 54 3/8" high, 27 7/8" wide. This next group of mirrors have details which can associate them with both the three dimensional Rococo and two-dimensional Classical periods; in fact they form a transition between the two styles. They retain the over-all fret-work outline and construction, while exhibiting decorative elements of classical derivation: open, loose finials and swags, flatter decoration, taller proportions, and wheat, oval, or urn motifs.

This mirror and the next, 436, have been associated with the merchant William Wilmerding whose labels have been found on similar mirrors. He sold mirrors in New York City from 1789 to about 1813. *Courtesy, The Henry Francis du Pont Winterthur Museum.*

436. Mahogany veneer and gilded wood, English or American, circa 1790. Like the preceeding mirror, this is associated with William Wilmerding's store in New York, by similar labeled mirrors (one is in the collection of the Museum of the City of New York). *Courtesy of David Stockwell, Inc.*

437. Mahogany veneer and gilded wood, American, circa 1790. A portrait bust in relief of George Washington decorates the crest while the top resembles architectural mirrors and the base resembles fret-work mirrors. The proportions are tall and lean, and a line of light stringing inlay is flanked by two gilded carved bands. *The White House Collection.*

435

436

437

438

440

439

438 and 439. Mahogany veneer and gilded wood, American, bearing the label of Nathan Ruggles of Hartford, Connecticut, circa 1810. The top crest of an architectural mirror, and sides and lower crest of a fret work mirror are combined with a purely Classical finial—vase with rather pedantic flowers. The label (439) is not so different from the mirror on which it is found. Nathan Ruggles (1774 to 1835) made and sold mirrors with Charles Mather, Junior from August, 1803 to July, 1804 when the firm Ruggles and Dunbar was formed with Azell Dunbar. This firm lasted until April, 1806 when Ruggles left to work by himself. He occupied several locations in Hartford and from 1818 to 1825 ran Ruggles Eagle Looking Glass Manufactory. *Courtesy, The Henry Francis du Pont Winterthur Museum.*

440. Mahogany and gilded wood, American, bearing the label of Nathan Ruggles of Hartford, Connecticut, circa 1810, 59" high, 28" wide. The beaded swags in the crests and gilded gesso Classical figure are very unusual decorations for a fret-work or architectural mirror. Ruggles did some improvising here. These Classical elements are so apparent, and coupled with the double gilded bands around the veneered frame, are a striking departure from the fret-work mirrors seen already in this book. *Courtesy Israel Sack, Inc., New York City.*

441. Mahogany veneer and gilded wood, American, circa 1810, 57" high, 22½" wide. The fret-work details are becoming more residual, less cut out design in the frame, very weak side swags, and a definite architectural crest with gilded caps. The carved phoenix is the most Rococo feature, which

443

442

441

seems to have very little relationship to the rest of the mirror. The oval patera has an inlaid pattern of a butterfly-a new motif for our story although other examples of it exist, and the stringing outline are Classical details.

442. Mahogany veneer on pine, and gilded wood, with spruce back board, American, probably Massachusetts, circa 1800, 59¾" high, 24 5/16" wide. The fret work details are loosing out to the Classical motifs here which give a light, fragile overall feeling to this mirror. The swags are quite weak, and finial large and light, while the scrolls of the pediment are pulled high and weak from the architectural roots where they derived. The eagle inlaid motif in the oval patera relates to examples

found in Massachusetts made furniture (see C. Montgomery. *American Furniture of the Federal Period.* inlay #96.) and can be compared with the eagle inlay on mirror 409. *Courtesy, The Henry Francis du Pont Winterthur Museum.*

443. Mahogany and gilded gesso, American, circa 1800, 53" high, 21¼" wide. Another American eagle insignia in oval patera is inlaid in the crest which has very high scrolls, oval rosettes, a Classical urn with thin, whispy flowers, and delicate side swags. The inlay and proportion are purely Classical as the decoration becomes flat rather than carved at this period. *Courtesy, The Diplomatic Reception Rooms, United States Department of State.*

444

445

444. **Mahogany on pine and gilded wood and gesso on wire, American, circa 1800.** The tall scrolls in the arch almost seem to grab this urn with drooping wheat and flowers. The oval patera inlay has a sea shell design. This decoration has become very flat, dispensing altogether with side swags or other carving. The trend is toward two dimensional design. *Courtesy of Herbert Schiffer Antiques, Inc.*
445 and 446. **Mahogany on pine and gilded wood with gesso and wire, American, bearing the label of B. Plain of New York, circa 1815.** The fret work outline is very restrained on this mirror where the decorative impact is wholly in the urn finial and flowers. The label (446) was used by Bartholomew Plain of New York in the 1815 to 1817 period when his businesss was at 33 Chatham Street. Plain was a gilder, mirror maker and merchant from 1802 until 1846 at various locations in New York City. The drawings of a dressing mirror and Masonic eye and compass may be clues of his work and membership. *Courtesy of Herbert Schiffer Antiques, Inc.*

446

447

448

447. Mahogany veneer and gilded gesso, American, bearing the label of James Stokes who sold it, circa 1810. The tightly scrolled volutes and pine-cone ears now associated with James Stokes' store and label (see mirrors 411 through 415) are combined with a Classical urn, flowers, and stringing inlay.

448. Mahogany veneer and gilded wood, American, Hudson River Valley, circa 1800, 60″ high, 22″ wide. This combination of transitional frame and eglomise panel has been associated with the Hudson River Valley area of New York and Connecticut. *Courtesy of David Stockwell, Inc.*

449

450

449. Mahogany and gilded wood, American, Hudson River Valley, circa 1800, 54″ high, 24″ wide. The eglomise panels in this group of Hudson River area transitional mirrors each have simple landscape with house decoration. *Courtesy of David Stockwell, Inc.*

450. Mahogany with gilded wood, American, Hudson River Valley, circa 1800, 72″ high, 27″ wide. This stately mirror is dripping with ornament of a delicate nature. The inlaid oval patera with American eagle design is associated with New England, and perhaps New York furniture, while the shell design in the lower crest is more generally found in New England. It is not known whether these details were imported from England where fine inlays were a craftsman's art, or whether they were made in America. Nevertheless, as assembled here, the mirror has an American personality. A related mirror in the collection of Sleepy Hollow Restoration, Tarrytown, New York has the following inscription stenciled on the back. "From Del Vecchio Looking Glass and Picture Frame Manufacturers, New York." (See C. Montgomery, *American Furniture of the Federal Period.* # 215.) *Courtesy, The Henry Francis du Pont Winterthur Museum.*

451

452

453

451. Gilded gesso and wood on marble plinth, English, circa 1770, 35" high, glass 23¼" diameter. These three unusual dressing mirrors (451, 452, 453) form a small group to illustrate the transition of this type of mirror from Rococo (451) to Classical (453).

The dramatic snakes (451) would have appealed to the reputed original owner of this mirror, David Garrick, the famous English eighteenth century actor, who is said to have used this at his villa in Hampton. It has an exotic design which seems closely related to Continental Rococo design. *Courtesy of the Victoria and Albert Museum.*

452. Walnut and walnut veneer, American, circa 1780 to 1795, 21 1/8" high, 11" diameter of tray, 9¼" wide at base. This dressing mirror resembles Queen Anne tables of the early part of the century, but the form is so unusual, it may be a unique piece. As the dressing stand is usually considered a later form, the date seems to be an intelligent guess. *Courtesy, The Henry Francis du Pont Winterthur Museum.*

453. Rosewood and brass, English, circa 1820, 62" high, glass 29" diameter. This elegant mirror would be a masterpiece at any age. It is enormous and fitted with a concave glass panel which magnifies its reflection. The contrasting stringing inlay around the glass helps to support an early nineteenth century date. The brass arm support is gilded, as are the flutings in the three columns in the support. The tripod base is carved in good Rococo leafy style. *Courtesy of the Victoria and Albert Museum.*

CLASSICAL

454

The group of mirrors which follow are designed in the Classical style. The overall characteristics are lightness, fragility, and motifs inspired by Greek and Roman art. This style can be seen as a reaction to the Rococo, or as the next generation's natural, swing-of-the-pendulum, preference for different and innovative designs. The style probably originated with Robert Adam who was a contemporary, although younger, of Thomas Chippendale, William Kent, and other Rococo designers, but his styles were not based on their origins. In 1754, Adam traveled to Rome where Piranesi was working, and studied the ruins of Classical antiquity first hand. Herulaneum and Pompeii were recently discovered and Adam was a keen observer and artistically influenced by this fresh, new approach. Back in England in 1758, he was able to convince a small but influential group of patrons who supported his ideas.

In comparison to William Kent's massive rooms and decorations, Robert Adam's work seemed "neat and pretty" to his critics (see Wills, page 33). In time, however, his concept was very much copied and popularized by leading designers, such as John Linnell and Matthias Lock.

456

455

457. **Gilded wood and plaster, French, from the Hotel de Tesse, Paris, late eighteenth century.** While Robert Adam was designing in the Classical style in England, a similar, ordered style was developing in France which became known as Louis XVI after the patron-King. In America, the English Classical style design books by George Hepplewhite and Robert Sheratan were very influential in creating the Federal style which lasted well into the nineteenth century. This French mirror is built into the room paneling and makes use of Classical designs in the moldings, floral wreath, and carefully bundled flowers joined by ropes at the sides. Every detail is balanced, and the columns and arch are angled to suggest deep perspective. *Courtesy of the Metropolitan Museum of Art.*

458. **Gilded wood, French, from the Pavilion de Marsan, Palais de Tuileries, Paris, late eighteenth century.** The French Louis XVI style is a clear interpretation of Classical motifs. *Courtesy of the Metropolitan Museum of Art. Gift of J. Pierpont Morgan, 1906.*

457

Gilded

454. **Gilded wood, English, circa 1770.** The use of leaves and branches on this mirror is a remnant of the Rococo style's pastoral derivation and use, while the design as a whole is light and Classical in feeling.

455. **Gilded pine and papier-mache, English, circa 1775, 92" high, 56" wide.** The profusion of bell-flower garlands and open design are Classical elements on this mirror. Some of the decoration is made of gilded papier-mache, an important media variation which enabled light and precise ornaments. *Courtesy of the Victoria and Albert Museum.*

456. **Gilded wood, English, circa 1760.** After designs by Robert Adam, the use of Antherium leaves, as in the arch of this mirror, became a standard decoration of the Classical style. The symmetrical and very rigidly ordered design is the antithesis of Rococo assymmetry being made at the same time in England. The two styles ran parallel for many years with the newer, Classical being further interpreted after the Rococo went out of fashion.

458

459

460

459. Gilded wood, Italian [Lombardy], circa 1770-1780. This mirror was made in Italy at the same time as the Classical Adam style was gaining acceptance in England. Many of the same garland and symmetrically organized elements are present, even though in this mirror the overall feeling is heavier. *Courtesy of the Metropolitan Museum of Art. Rogers Fund, 1923.*

460. Gilded wood and gesso, English, designed by Robert Adam, circa 1760. Robert Adam designed this mirror as one of a pair in the ballroom at Syon House. It can be considered a source for so many details in this group and as setting a precedent for the style. The Classical figures reflected in the mirror are derived directly from ancient Italian art. The drawing for it is currently in the Sloan Museum. *Courtesy, Syon House, Brentford, Middlesex, Great Britain.*

461

462

461. Gilded wood, English, circa 1770.
462 and 463. Gilded white pine and adler, and gesso over wire, American, bearing the label of James Todd of Portland, Maine, circa 1825, 49" high, 22¼" wide. The gilded gesso flowers and leaves are quite delicately made and arranged while the urn appears two large for the glass panel from this perspective. The label (463) shows a frame of the very latest design for this period circa 1825, even though the mirror itself is of an earlier style. James Todd (1795-1884) used this and four other similar labels while he worked on Exchange Street in Portland, Maine between 1820 and 1831. Thereafter, he worked in partnership with Samuel Beckett at other locations in Portland until 1882.

463

464. Drawings of pier glasses, English, by George Hepplewhite, *"The Cabinet-Maker and Upholsterer's Guide,"* **3rd edition, 1794.** The design books published by George Hepplewhite were very influential in establishing the Classical style's popularity, especially in America. Details from these drawings will be seen on many mirrors in this section.

London, Published Sept.ʳ 1ˢᵗ 1787, by I. & J. Taylor, N.ᵒ 56, High Holborn.

465

466

465. Gilded wood, English, circa 1780. The frame is quite restrained on this mirror. Only the crest decoration gives the delicate, fragile appearance to the mirror.

466. Gilded wood, gesso and wire, English, circa 1770 to 1775.

467. Gilded wood, designed by Robert Adam and made by William Ince and John Mayhew, billed October 5, 1769, 105" high, 66" wide. After Robert Adam's design, this mirror was made by William Ince and John Mayhew in London for the Tapestry Room at Croome Court, Worcestershire. A bill from Ince and Mayhew dated October 5, 1769 to the Earl of Coventry for £35 describes this mirror exactly. Ince and Mayhew had a long lasting partnership from 1759 until 1783 as cabinetmakers. They published *The Universal System of Household Furniture* as Rococo design leaflets between 1759 and 1762 to compete with Thomas Chippendale's *Gentlemen's and Cabinet-Maker's Director.* This mirror has nothing to do with those Rococo designs, but rather is the work of the cabinetmakers under Robert Adam's direction and Classical design. *Courtesy of the Metropolitan Museum of Art. Gift of the Samuel H. Kress Foundation.*

467

468

469

468. Drawing for a mirror, English, by Thomas Sheraton, circa 1793. Thomas Sheraton (1751-1806) was a cabinetmaker and book publisher of the current taste who published a series of designs from 1791 to 1793 which he collected into a volume in 1793, *The Cabinetmaker and Upholsterer's Drawing Book.* In 1803, *The Cabinet Dictionary* was published, and from 1804 until 1806, parts of *The Cabinetmaker, Upholsterer's and General Artist's Encyclopedia* were released. It is not known how much, if any, of the designs were his own or if they were drawings of existing furniture in the London work shops, which is more likely. His tradecard states that he "Teaches Perspective, Architecture and Ornaments, makes designs for cabinet-makers, and sells all kinds of Drawing books, etc." Sheraton stated in his *Drawing Book* "I have made it my business to apply to the best workmen in different shops, to obtain their assistance in the explanation of such pieces as they have been most acquainted with." (London: T. Bensley, 1802, pages 352-353.) His books were very influential, especially among American cabinetmakers of the next generation. *Courtesy of the Victoria and Albert Museum.*

470

471

472

473

469 and 470. Gilded wood, English, circa 1774. This mirror is shown twice, once in its correct over-mantle setting, and again closer to emphasize the carved detail. It is a masterpiece of carved detail and Classical style, altogether an ambitious concept executed well. The Classical griffins, urns, and bell flowers remain in balanced scale with the overall design. It was made for the home Bradbourne in Kent. *Courtesy of the Victoria and Albert Museum.*

471. Gilded wood, gesso and wire, English, circa 1780.

472. Gilded wood and paneling, French, late eighteenth century. The mirror was built into the room paneling for the Hotel de Cabris, Grasse. (A hotel then meant a great town house.) The Classical ornamentation extends well above the mirrored glass into the surrounding paneling. *Courtesy of the Metropolitan Museum of Art.*

473. Gilded wood, English, circa 1790. The shape is quite unusual here, yet the many decorative motifs, including ram's heads, are typical Classical details and the overall design is well within the prevailing taste.

474

475

474. Drawing, English, by Robert Adam, circa 1770, 1½″ wide, 9½″ high. Robert Adam drew this mirror and table design for Robert Child's house on Berkeley Square, London. The drawing is retained at Osterley Park. *Courtesy of the Victoria and Albert Museum.*

475. Drawing, English, by William and John Linnell, circa 1770, 10 7/8″ high, 7″ wide. The Linnell firm designed this Classical pier mirror and table for the drawing room at Shardeloes for William Drake, Esquire. Each half is designed independently of the other, yet an overall balance of design is achieved. An alternate crest design was shown on an attached flap. *Courtesy of the Victoria and Albert Museum.*

476. Carved and gilded wood and molded gesso, English, circa 1780. The medallion in the crest is a charming addition which continues the blythe feeling of the decoration. The use of two small panels of glass is unusual at this period since long glass was readily available in London.

477. Drawing, English, signed J. Carter, circa 1776, 7 5/8″ high, 5 5/8″ wide. John Carter designed this girandole in the Classical style as plate 82 in *The Builders' Magazine* 1774-1786. The trapezoid shape of the glass is unusual. The decoration is fanciful yet balanced. *Courtesy of the Victoria and Albert Museum.*

476

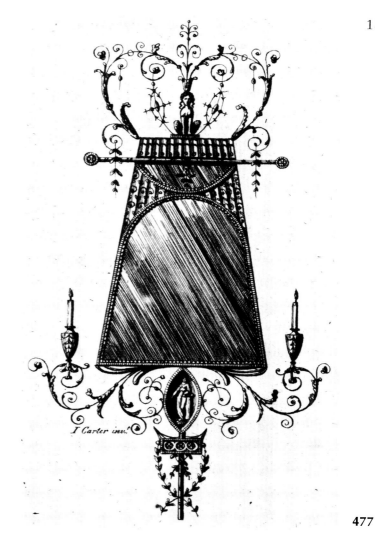

I Carter inv.

477

478

478. Gilded wood and brass, English or American, circa 1780. This is a more simple version of the preceeding mirror (477) lacking all ornamentation. The mirror is interesting in design although mediocre in detail. *Courtesy of Herbert Schiffer Antiques, Inc.*

Wooden

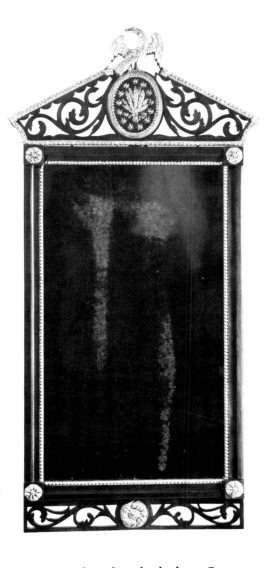

481

482

479

480

479. Fruitwood and etched glass, European, circ[a] 1800, 49" high, 18" wide. This type of mirror wa[s] made from Italy to Denmark. The etched uppe[r] panel and placement of the stringing inlay ar[e] peculiar to this style. (This mirror is shown as plat[e] 3144 in Wallace Nutting's *Furniture Treasury*, volume II, and identified as American, but th[e] author disagrees.) *Courtesy of Alfred Bullard, Inc[.]*

480. Mahogany veneer on pine, Danish, circa 178[5,] 66" high, 25" wide. The Classical elements ar[e] handled with a relatively heavy hand in this mirro[r,] probably from the Copenhagen region of Denmark[.] In 1784, American merchants advertised importe[d] looking glasses from England, France, Holland an[d] Germany. (Charles Montgomery, *American Furn[i-] ture of the Federal Period,* page 354.) From th[is] period forward, design ideas traveled quickly a[ll] over the world and influenced diverse style[s.] *Courtesy of James Billings, London.*

481. Gilded wood and marble veneer, Spanish o[r] Portuguese, circa 1790 to 1810, 41" high, 21" wide. This type of mirror has been called Bilboa since the[y] are thought to have come from the port of Bilboa i[n] Spain near the French border, a stopping place fo[r] American ships. Many mirrors of this type are foun[d] in American homes.

The frame is veneered with marble, and a paint-ing rests in the oval medallion of the crest. *Courtesy of the Metropolitan Museum of Art. Rogers Fund, 1919.*

482. Mahogany and gilded wood, American, attributed to Constant Bailey of Newport, Rhode Island, circa 1790-1800, 49" high, 20¼" wide. The straight rectangular shape and pierced decoration are generally considered German interpretations of this Classical style (see 483), yet the eagle and fifteen stars are American motifs. It is quite a[n]

483

484

unusual mirror for either location. This mirror is attributed to the cabinetmaker Constant Bailey who worked for the firm Goddard and Townsend of Newport. It descended in his family and was originally one of a pair. *Courtesy the Diplomatic Reception Rooms, United States Department of State.*

483. Mahogany with brass mounts and eglomise, German, circa 1810. German mirrors were imported to America through the nineteenth century, and this is a good example of the straight rectangular form which most of them have. The eglomise panel and cast brass mounts are typical of their ornamentation. Several of this type are shown in Wallace Nutting's *Furniture Treasury*, indicating how long they had been associated with American decoration. *Courtesy of David Stockwell, Inc.*

484. Mahogany veneer and gilded wood, Irish, circa 1780. The combination of a plain molded mahogany frame and gilded crests is unusual but quite a successful design. A brass candle arm was fitted to the holder in the lower crest. *Courtesy of M. Turpin Antiques.*

485. All glass, Irish, circa 1780. The successful glass industry in Ireland, especially at Waterford, led to the design of all-glass decorations from that source for the British interiors. The border of cut glass pieces (like jewels) was an Irish characteristic, and the use of blue glass was popular throughout this period. *Courtesy of the Victoria and Albert Museum.*

485

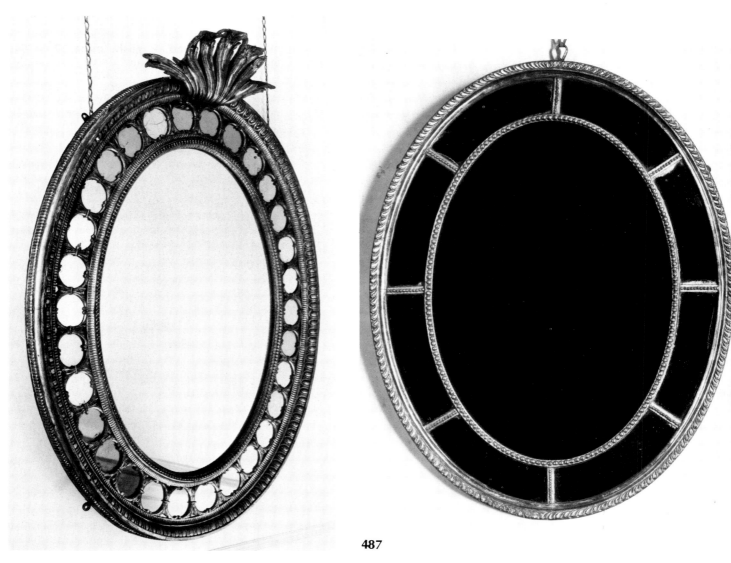

486

487

488

486. Gilded wood, English, circa 1795, 54" high, 39½" wide. The Classical style is handled with heavier moldings on this example, and the carving on top is a remnant of Rococo design.

487. Gilded wood, English, circa 1795, 27" high, 21" wide. Void of nearly all decorative ornament, this mirror's simplicity is its best feature. By using only the Classical oval shape and glass borders, the style is achieved directly.

488. Gilded wood, English, circa 1770, 51" wide, 42" high. The fan-shaped carvings in the border relate to fan inlay on some Classical style furniture. *Courtesy of Alfred Bullard, Inc.*

489. Gilded wood and eglomise glass, English, circa 1775. This border of eglomise glass has eight carved, wooden, oval patera evenly spaced between the beaded and rope-carved borders. The overall design is relatively heavy because of the border.

490. Gilded wood, French, 1798, 35½" high, 29" wide. This is one of a set of three mirrors bought by Stephen Girard, merchant of Philadelphia, from Bordeaux, France in 1798. The design is simple and the use of a carved ribbon motif in the border was more common in France, but used throughout Europe on Classical mirrors. *The Stephen Girard Collection, Girard College, Philadelphia.*

491. Gilded wood, French, late eighteenth century. The ribbon design is used again in this border, with the additional Classical details of laurel leaves and tassels in the lower crest, and the heraldic arms of the French King and crown in the upper crest. The inference that the French King was to France as the Roman Emperor was to his empire is clearly implied. Such artistic suggestion was not lost when Napoleon assumed power; he loved this sort of subtle significance. *Courtesy of the Metropolitan Museum of Art. Gift of J. Pierpont Morgan, 1906.*

489

490

491

492

493

492. Painted wood, French, late eighteenth century. This is a lovely example of Classical design with the finest quality carving. The ribbon looks so fragile and the flowers are truly life-like. While the design conveys a perishable quality, the use of grey paint on the surface simulating stone gives a solid appearance to this mirror. *Courtesy of the Metropolitan Museum of Art, gift of J. Pierpont Morgan, 1906 [07.225.437]*

493. Gilded wood and wire, probably American, circa 1780. During the eighteenth century, the pineapple was a symbol for hospitality, and here it is incorporated twice into the Classical design of this mirror. *Courtesy of Herbert Schiffer Antiques.*

494. Gilded wood, English or American, circa 1790, 36½" high, 17 1/8" wide. This is the first example in this book of an eagle holding a chain with gilded balls in its mouth—a motif which was used often on convex mirrors in the early nineteenth century (see 608, 612, and 613.) The Classical oval shape is nicely combined with this little bit of ornamentation. *Courtesy of Herbert Schiffer Antiques.*

494

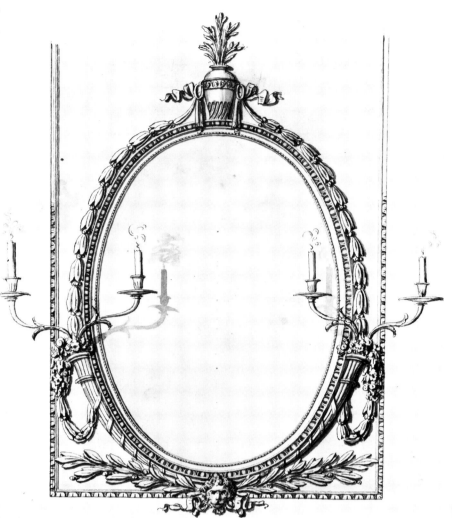

495

495. Drawing, English, by John Yenn, circa 1780. This is quite an attractive design using many Classical elements. The border design seems to indicate that it was meant to represent a built-in wall panel with set-in mirrored glass and extending candle arms. *Courtesy of the Victoria and Albert Museum.*
496. Gilded wood, English, circa 1780.
497. Gilded wood, English, circa 1780. These mirrors are quite light in feeling because of their small motifs and very open design.

496

497

498

499

500

501

498. Gilded wood, English, circa 1770. The oval patera in the border resemble those on mirror 489 but this mirror has a lighter overall feeling because the decorations extend the the viewer's vision out from the oval form. *Courtesy of Alfred Bullard, Inc.*

499. Gilded wood, wire and gesso, English, circa 1775, 35" high, 17" wide. *Courtesy of Alfred Bullard, Inc.*

500. Gilded wood, English, circa 1780, 58" high, 34" wide. *Courtesy of Alfred Bullard, Inc.*

501. Gilded wood, English, circa 1780, 41" high, 20" wide.

502. Gilded wood, English, circa 1780. The lyre-shaped motif in the crest is unusual on mirrors, but was one of the Classical style motifs used on furniture, particularly American tables from Duncan Phyfe's New York workshop, for example.

503. Gilded wood, English, circa 1780. The grape vine motif was a late development of the Classical style, and has caused some to consider the period of its use "The finicky period." The elements are tiny and busy, perhaps shimmering with tendrils extending into void space. Call it refinement or fussy.

504. Gilded wood, English, circa 1780, 64½" high, 31" wide.

502

503

504

505

506

507

505. Walnut veneer and gilding, English or American, circa 1780, 30″ high, 16″ wide. It is unusual to find walnut veneer being used at this late date. The oval shape is very unusual for the fret-work form; that and the use of stringing inlay place this as a transitional mirror between Rococo and Classical styles. Damaged areas to the bird's head and scrolls of the frame are noted. *Courtesy of Alfred Bullard, Inc.*

506. Mahogany veneer and gilded wood, English or American, circa 1780. Helen Comstock identifies this style mirror as coming from the New York area (see *The Looking Glass in America, 1700-1825,* pages 63 to 65.) The details on this mirror (feathers and gilded caps) are more likely of English origin as they relate closely with the fretwork group (314 to 330) in this book. The oval shape and use of stringing inlay place its design into the Classical vocabulary. *Courtesy of Samuel Schwartz.*

507. Mahogany veneer and gilded wood, English or American, circa 1780, 45″ high, 20½″ wide. The delicate pierced decoration in the crests, combined with the gilding and phoenix carving, highly suggest an English origin for this mirror. For all its Rococo details, its oval shape seems to place it in the transition period toward the Classical style. *Courtesy of the Metropolitan Museum of Art. Sansbury-Mills Fund, 1952.*

508

509

508. Gilded wood, American, circa 1790, 58" high, 32" wide. The shape of this mirror closely resembles the eagle insignia adopted by the Society of the Cincinnati after the American Revolutionary War. The national motto is lettered on the ribbon in the eagle's mouth. This may be a unique mirror design—it certainly ranks among the finer American folk art carvings. *Courtesy of Israel Sack, Inc., New York City.*

509. Gilded wood, American, mid-nineteenth century. This mirror is one of a pair portraying the cornucopia as a motif of the mid-nineteenth century. The ovoid shape is a liberal interpretation of the Classical oval form, and the fruit decoration became a favorite ornament about 1830. The gilding is done in two tones—the darker areas burnished for contrast. *Courtesy of Craig and Tarlton, Inc.*

510. Gilded wood and gesso, American, circa 1810 to 1820, 58" high, 36½" wide. This unusual mirror relates to the Classical style in its oval shape, the Empire style in its use of cornucopia and acorns, and convex girandoles in its design. It has been placed here as a late example of the Classical style, being a transitional example between Classical and Empire. The carving is fair, but dominated by the gesso and gilding. It is a most interesting combination of details, and the Albany area of New York has been suggested as a possible place of origin. *Courtesy of Israel Sack, Inc., New York City.*

510

511. Gilded wood, French, circa 1770. This room abounds with Classical motifs in every form of decoration. Mirrors are used over the mantle, in the doors, and high arches. This is the boudoir of Madame de Scully, maid of honor to Marie Antoinette and wife of the army Paymaster General. *Courtesy of the Victoria and Albert Museum.*

512

511

512 and 513. Pair of trumeau paintings [paintings over mirrors] in gilded wood frames, French, circa 1790, 62½" high, 33½" wide. Wealthy merchant Stephen Girard, originally French but settled in Philadelphia, ordered this pair of trumeau paintings from France about 1790 for his home. The scenes are romantic fantasies with a prettiness of Classical sculpture. The frames are only structural, with little attempt of being decorative. *The Steven Girard Collection. Girard College, Philadelphia.*

514. Gilded wood, Italian, Lombardy, circa 1780. The Classical elements in the gilded decorations of this cabinet or closet create a "horror-vaccui", (dread of the undecorated,) which is not relieved by the plain mirrored panels, but only made more so by the reflections. There are eight pairs of doors and eight mirrored sections in this room. The frame is a simple running molding of repeated beads. *Courtesy of the Victoria and Albert Museum.*

513

514

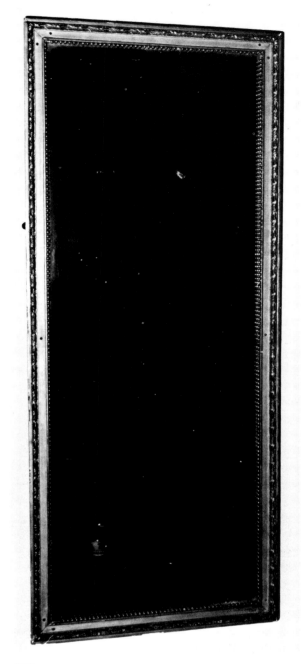

515

516

515. Gilded wood, French, circa 1810, 73" high, 30" wide. *The Stephen Girard Collection. Girard College, Philadelphia.*

516. Gilded wood, English, circa 1775, 66" high, 48" wide. The beading and oval patera introduced by Robert Adam in the 1760's was used by later Classical designers to create delicate and light mirrors of this overmantle type by 1775. The decoration is becoming less important than the function and architectural merits of the design.

517. Watercolor drawing, English, circa 1800. This simple mirror was designed for Bletchworth Castle in England about 1800. The molding cross section and detail show the outline of the carving and use of small medallions symmetrically placed on the frame. Balance and order were always controlled in the Classical style. *Courtesy of the Victoria and Albert Museum.*

518. Gilded wood and eglomise, English and Chinese, circa 1785. The reverse painted glass panel was decorated in China and framed in England with this simple molded and gilded border. During this period, Oriental fantasies continued to be popular, such as those at the Royal Pavilion at Brighton. *Courtesy of M. Turpin Antiques.*

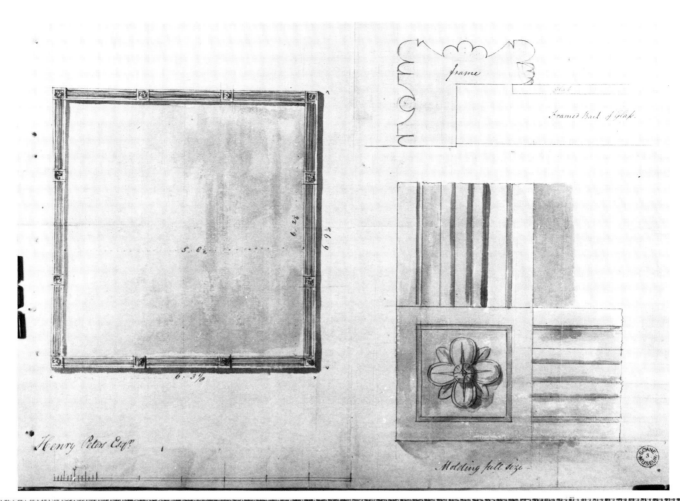

Henry Peters Esq.

Molding full size

517

518

519

520

**519. Painted wood, American, circa 1795, 23 7/8"
high, 15 7/8" wide, 10 1/8" deep.** Dressing mirrors
became a popular form of furniture in the late
Classical period, and were included among the
drawings of George Hepplewhite's *The Cabinet-
Maker and Upholsterer's Guide,* published 1788,
1789, and 1794. This book was published by the
author's widow Alice two years after his death, and
has been one of the most influential design books
of that century, probably second to Thomas Chippendale's. The Hepplewhite designs are of somewhat
questionable origin, not that they were not probably drawn by him, but that not a single piece of
furniture by his hand has survived, and the probate
of his estate proved only a very modest amount.
This was not a wildly successful or prolific designer
until the book was published, apparently. Perhaps
he, like Thomas Sheraton, drew furniture existing in
London cabinetshops at the time (see 468) rather
than creating these designs from his own imagination. The book, however, established his name for-
ever associated with the late Classical style, and
especially in America is the root of much furniture.

Dressing mirror 519 was made in America,
possibly in New York. The ball and claw feet are
remnants of the Rococo style as it was interpreted in New York, while the rest of the piece is
Classical in style. There are only two drawers in the
case. For the sake of Classical symmetry, the front is
made to look like three drawers, but the left two are
all one. *Courtesy, The Henry Francis du Pont
Winterthur Museum.*

521. Mahogany and Satinwood, English, circa 1770.
The cabinetwork on this dressing mirror is of the
highest achievement and most elegant design. It is
very unusual to find cut work in glass at this period,
yet here it is. The frame and case are inlaid in a
most interesting pattern including stringing, bell
flowers, and oval patera with contrasting wood
panels. This is a joy to observe. The quality was
repeated on the best tea caddies of this period.
Courtesy of David Stockwell, Inc.

**520. Rosewood veneer, English, circa 1795, 24 3/8"
high, 21" wide, 9½" deep.** This design is called a
"skeleton" mirror. The oval adjustable frame pivots
within the supporting frame. *Courtesy of the
Victoria and Albert Museum.*

524. Mahogany, American, circa 1770-1790, 27 1/8" high, 15½" wide, 8" deep. The case of this dressing mirror is made in the Rococo style of New England with the serpentined blocking shape on ball and claw feet. The mirror, in contrast, is shield shaped and outlined in the late Classical style. *Courtesy, The Henry Francis du Pont Winterthur Museum.*

522

tephen Badlam, jun

No. 42, CORNHILL,

BOSTON:

A CONSTANT SUPPLY OF FASHIONABLE

LOOKING-GLASSES,

Wholesale & Retail.

523

522 and 523. Mahogany, American, bearing the label of Stephen Badlam, Junior, of Boston, circa 1801 to 1804, 20 5/16" high, 20 7/8" wide, 10" deep. In style, this is a good example of the late Classical use of wood grain and inlay in a two dimensional scheme as decoration rather than any carving or applied ornament. Especially in America this preference was true in the earliest years of the nineteenth century. Also significant is the label fixed to the back of this dressing mirror. Stephen Badlam, Junior (1779-1847) was the son of the prominent cabinet and mirror maker Stephen Badlam, Senior (1751 to 1815). In 1801, Stephen, Junior, was half of the partnership Howe and Badlam with his brother-in-law James Blake Howe (1773 to 1844) which advertised looking glasses for sale. This partnership seems to have lasted until 1803 when Stephen Badlam, Junior advertised alone. No other record of his work is listed in city directories, and he left for Buenos Aires in 1811. It is not known if he sold his own, his father's, or other craftsmen's mirrors. *Courtesy, The Henry Francis du Pont Winterthur Museum.*

524

527 and 528. Mahogany, American, bearing the label of Thomas Natt of Philadelphia, circa 1823. This is a typical representative of the American late Classical style dressing mirror. Thomas Natt worked as a carver, gilder, looking-glass maker, importer and merchant in Philadelphia between 1809 and 1841 when his son Thomas J. Natt joined him and eventually took over the firm. From 1810 until 1819, his shop is located in Philadelphia directories at 76 Chestnut Street, the address on the label. Other labels are known (see Betty Ring, "Check list of looking-glass and frame makers and merchants known by their labels", *The Magazine Antiques*, May, 1981, page 1188) with addresses in Philadelphia at locations recorded later in his career. *Courtesy of Herbert Schiffer Antiques.*

527

526

525 and 526. Mahogany, white pine and tulip poplar, American, bearing the label of Charles Del Vecchio of New York, circa 1835, 16" high, 14¾" wide, 6 3/8" deep. The dressing glass is as plain in design as it can be, which lets the attractive grain of the wood assume all of the decorative features. Charles Del Vecchio (circa 1786-1854) ran a mirror and print store in New York from 1812 until 1853 at various locations and with interruptions. In 1828 he first is listed as a looking glass manufacturer. The label (526) has the address 44 Chatham Street where he was between 1831 and 1840. From 1841 he advertises with his son, Charles, Junior. He was one of the brothers (see 405 and 405) who came to America in 1800 from Moltrasio, Italy. *Courtesy, The Henry Francis du Pont Winterthur Museum.*

528

529

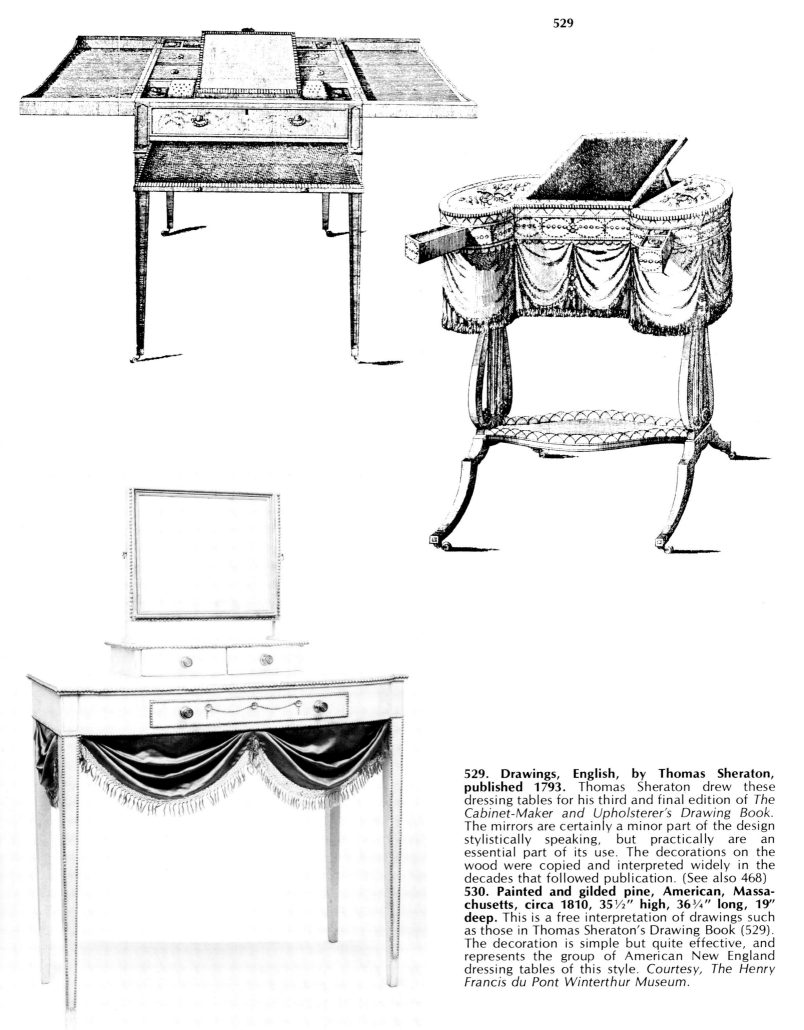

529. Drawings, English, by Thomas Sheraton, published 1793. Thomas Sheraton drew these dressing tables for his third and final edition of *The Cabinet-Maker and Upholsterer's Drawing Book.* The mirrors are certainly a minor part of the design stylistically speaking, but practically are an essential part of its use. The decorations on the wood were copied and interpreted widely in the decades that followed publication. (See also 468)

530. Painted and gilded pine, American, Massachusetts, circa 1810, 35½" high, 36¾" long, 19" deep. This is a free interpretation of drawings such as those in Thomas Sheraton's Drawing Book (529). The decoration is simple but quite effective, and represents the group of American New England dressing tables of this style. *Courtesy, The Henry Francis du Pont Winterthur Museum.*

530

531

532

531. Mahogany, American, circa 1800. The mirrors in the raised back board and interior compartment of this fine quality lady's writing desk give the illustion of depth to the design. It is a detail found occasionally on the better Classical desks from Philadelphia and Baltimore at this period. *Courtesy of the Philadelphia Museum of Art. Bequest of Fannie Norris.*

532. Lacquered wood, Chinese, early nineteenth century. Chinese lacquer work was imported to England and America between about 1810 and 1850 on the same merchant ships that brought tea, silk and porcelain. This is a particularly fine and well preserved dressing mirror with gilt patterned decoration on the European Classical style frame and Chinese contoured base. *Courtesy of David Stockwell Antiques.*

533. Lacquered wood, Chinese, circa 1810. The placement of Classical shaped dressing mirror on an over-sized stepped case and separate stand is strictly a Chinese innovation which was intended for export trade. The lacquer work decoration is marvelous. *Courtesy of The Museum of the American China Trade.*

533

534

535

536

534. Lacquered wood, Chinese, circa 1820. This is a superb example of the style in both design and decoration. *Courtesy of The Museum of the American China Trade.*

535. Mahogany on tulip poplar and chestnut, American, by Joseph Rawson & Son of Providence, Rhode Island, circa 1825, 72½" high, 38" wide, 21" deep. The dressing table form evolved from "lowboys" in the eighteenth century to chests of drawers with attached mirrors in the nineteenth century. This is a superb design using both carved and inlaid decorations. The lower skirt band is composed of demi-lunettes once thought to be almost a signature detail of the workshop of Thomas Seymour of Salem, Massachusetts. However, here it is on this dressing table which was made for Mary Thomas Rivers by Joseph Rawson & Sons of Providence, Rhode Island. Joseph Rawson (d. 1835) was the son of cabinetmaker Grindall Rawson (1719-1803) of Providence. His listing on his own in the city directory begins in 1824 and by 1832, his three sons George B., Samuel, and Joseph, Junior are also mentioned. *Courtesy of David Stockwell, Inc.*

536. Mahogany and maple veneer, American, New England, probably Boston, circa 1820, 71 1/8" high, 38 1/8" wide, 21¼" deep. The design of these three (535, 536, 537) dressing tables is remarkable similar, strongly suggesting a common design source from a published book or designer. *Courtesy, The Henry Francis du Pont Winterthur Museum.*

537

538

539

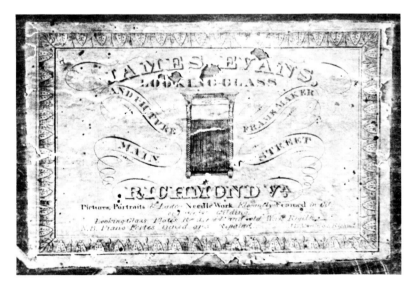

540

**537 and 538. Mahogany and mahogany veneer,
American, bearing the label of Levi Ruggles of
Boston, circa 1815, 75″ high, 38½″ wide, 23¼″
deep.** Levi Ruggles is listed in the Boston city
directory for 1810, and the label (538) gives the
address at No. 2 Winter Street, Boston. Nothing
more is known of his work. *Courtesy, The Henry
Francis du Pont Winterthur Museum.*

**539 and 540. Mahogany and white pine, American,
bearing the label of James Evans of Richmond,
Virginia, circa 1820, 25 1/8″ high, 23″ wide, 7¾″
deep.** James Evans was a carver, gilder and mirror
and frame maker in New York City from 1806 to
1816 before he moved to Richmond, Virginia where
he continued his work until about 1830. The label
(540) gives his address as Main Street, and also lists
"Piano Fortes tuned and Repaired." *Courtesy the
Museum of Early Southern Decorative Arts,
Winston, Salem, North Carolina.*

**541 and 542. Mahogany and birch, American,
probably New York City, 1817, 34 3/8″ high, 26″
wide.** The four small brass feet attached and
probably original to this dressing mirror are the first
appearance in this book of these details which
became stock items of the Empire style which
follows. They are attached to an otherwide late
Classical mirror, and thus begins the transitional
style as shown here.

The mirror has survived with its original receipt
from the merchants Barelli, Torre & Co., of
Charleston, South Carolina who sold it to Mr. Ker
Boyce/Boyce & Johnston, December 24, 1817 for
$22. A Christmas present? *Courtesy Museum of
Early Southern Decorative Arts, Winston, Salem,
North Carolina.*

541

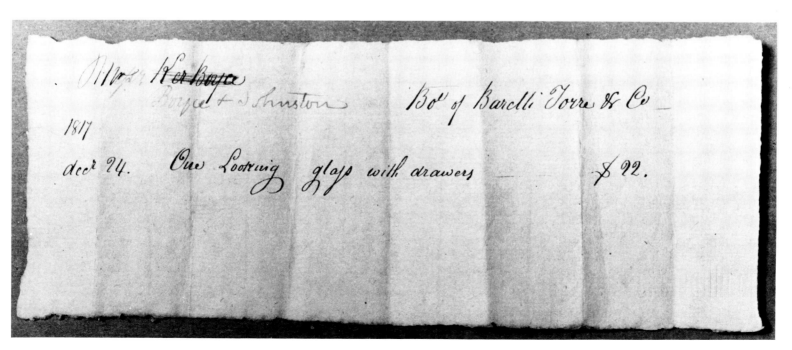

542

543

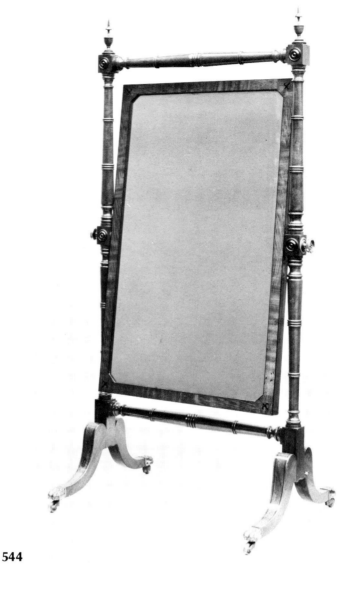

544

CHARLES N. ROBINSON,

CARVER AND GILDER,

Manufactures, and has Constantly on hand, at his LOOKING-GLASS a

PICTURE STORE,

NO. 56, SOUTH SECOND STREET,

PHILA'

A GENERAL ASSORT.

Gilt Framed Looking-Glasses,   ait Frames,
Mahogany   do.   do.   Girandoles,
Toilet   do.   do.   Brackets,
Pocket   do.   do.   Bed and Window Cor
Miniature Frames,

With a great variety of Profile Frames, WHOLESALE *and* RETAI

Looking Glasses, Needle Works and Pictures, elegantly Framed.

N. B.   Looking-G   and New Quicksilvere

543a

543 and 543a. Mahogany, American, bearing the label of Charles N. Robinson of Philadelphia, circa 1815, 37¼" high, 23" wide. Standing mirrors are sometimes called "cheval" (French for horse) mirrors because they stand on four legs like a horse. They are more common in the nineteenth century, but grew (also in size) out of the dressing glass tradition. Charles N. Robinson was a carver, gilder and mirror maker in Philadelphia between 1811 and 1857. This label (543a) lists his address at No. 56 South Second Street where he carried on his business between 1813 and 1818. *Courtesy, The Henry Francis du Pont Winterthur Museum.*

544. Mahogany, English, circa 1820. The use here of ring turnings, ebony inlay, and brass paw feet signal the beginning of a new style which emerged from the French Regency period. *Courtesy of the Victoria and Albert Museum.*

545 and 546. Mahogany, American, bearing the label of William S. Reddin of New York, circa 1830. As in the previous mirror, turned elements signal a shift in style, but this is also a very unusual form. It is considered a type of cheval mirror for dressing, but the pedestal and tripod base have little precedent, (see also 393 and 452). William S. Reddin was a mirror and frame maker in New York City between 1828 and 1837 working at 103 William Street. *Courtesy, Frank Schwarz & Son. Philadelphia.*

547. Amboyna veneer and bronze ormolu, French, attributed to the Jacob family of Paris, circa 1820. The Classical decorations are applied to this cheval

545

546

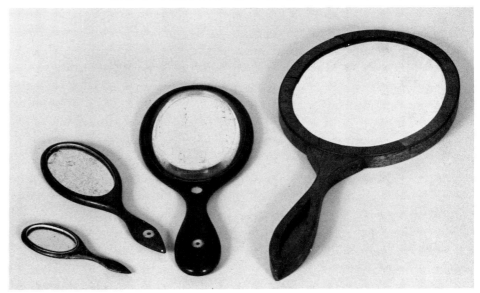

548

mirror in gilded bronze pieces over highly figured amboyna wood veneer. The overall design is merely a support for the decorations, which are beautiful.

George Jacob (1739-1814) was the celebrated head of a very successful family of French cabinet makers. He was extremely creative in the emerging Classical style, and made furniture of novel ideas for Marie Antoinette. His sons Georges and Francois-Honore continued his business from 1796. After son Georges died, the father and Francois-Honore formed Jacob Desmalter & Cie (the Desmalter from a small estate they owned in Burgundy). Francois was a brilliant designer and

craftsman who ran an enormous business with fifteen workshops. His son Georges-Alphonse Jacob-Desmalter (b. 1799) continued the business until 1847. *Courtesy of The Metropolitan Museum of Art. Fletcher Fund, 1924.*

548. Mahogany, English or American, circa 1800. Largest 13½″ long and 6″ wide; Smallest 4″ long and 1¼″ wide. This group of hand mirrors represents a type which was made in large numbers and various sizes. *Courtesy of Herbert Schiffer Antiques.*

549. Gilded wood and eglomise, American, probably New York City or Albany, circa 1800. The use of vertical pillars flanking an eglomise and a mirrored panel began about 1800 in America. Some people call this style "tabernacle", but we have found no reference to that term in references of the period. The frame sometimes incorporated applied balls and other decorative Classical style elements. This example (549) is part of a group associated with Albany, New York or New York City, with free-standing ornaments on top of the frame. The oval decorated panel in the crest has a cupid's bow and arrows and torch of eternal devotion suggesting its original purpose as a wedding gift. *Courtesy of David Stockwell, Inc.*

550. Gilded wood and gesso, American, probably New York City or Albany, circa 1800. If the previous mirror suggests nuptial bliss, the decoration of this mirror proclaims it with the same motifs enlarged and in three dimensions. The overall character of the two mirrors is comparable, and the pillars are similar. *The White House Collection.*

551. Gilded wood and gesso, and eglomise, American, circa 1800, 56" high, 27" wide. The eglomise panel is painted on two layers of glass to add depth to the design. The frame supports acorn and oak leaf carving of fine detail, and the double pillars at the sides display the Classical origins of this design. *Courtesy of Israel Sack, Inc., New York City.*

Pillars

549

550

551

552 and 553. Gilded wood and gesso, eglomise, American, circa 1800, 60" high, 29" wide. George Washington's death on December 14, 1799 touched off an American national fashion for mourning which lasted for several years. Embroidered mourning pictures, needlework samplers with mourning scenes, mourning jewelery and domestic decorative arts were made popular. The Classical motifs of urns and swags which were already part of the artistic vocabulary at this time merely were interpreted as reliquary urns and mourning swags. This mirror displays beaded swags, weeping willows, contained flags, stacked weapons, and a reliquary urn bearing a printed paper bust of George Washington. In case these subtle devices do not adequately convey the message, the plinth bears the inscription (553) which leaves no doubt unsatisfied. *Courtesy of M. Finkel & Daughter.*

554. Gilded wood, and eglomise, American, circa 1802, 37½" high, 19" wide. The eglomise decoration of this mirror continues the story of fashionable mourning (see 553), but here both George and Martha (died 1802) Washington are shown in portraiture on the tomb. Mount Vernon is depicted in the background, and a group of fashionably dressed people are shown paying their respects to the National heros. The frame incorporates incised and painted geometric decorations. *Courtesy, The Henry Francis du Pont Winterthur Museum.*

552

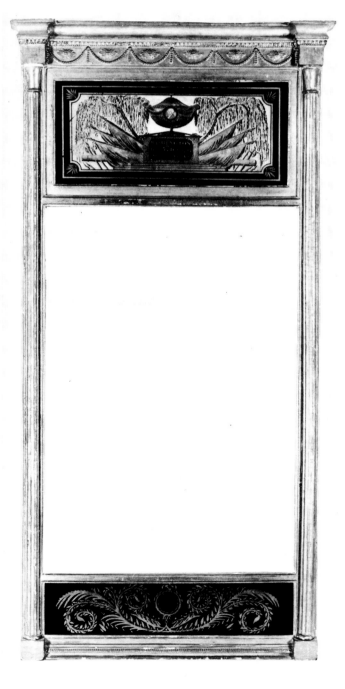

554

553

555

556

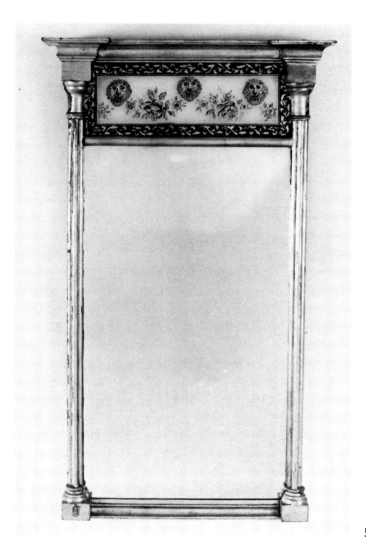

557

558

559

560

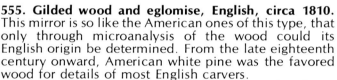

555. Gilded wood and eglomise, English, circa 1810.
This mirror is so like the American ones of this type, that only through microanalysis of the wood could its English origin be determined. From the late eighteenth century onward, American white pine was the favored wood for details of most English carvers.

556. Gilded wood and eglomise, American, bearing a label of Edward Lothrop of Boston, circa 1820, 31½" high, 19¾" wide. The eglomise glass is decorated with a stencilled border (similar to stencilled furniture of this period) around the free hand landscape. Another Lothrop mirror is shown (428) with its label (429). *Courtesy of the Honolulu Academy of Arts, given in memory of her husband by Mrs. Lee Offutt, 1974.*

557. Gilded wood and eglomise, English or American, circa 1815. The amusing lion's heads holding rings in the eglomise are copies of brass and fire gilt furniture handles used at this time.

558. Gilded wood and eglomise, English or American, circa 1815. This military theme on the eglomise glass is very unusual. Landscapes and naval scenes were more common.

559. Gilded wood, gesso, and eglomise, English, circa 1810. The molded gesso decorations on this mirror are very carefully made with fine detail. Some of the companies that made and supplied gesso ornaments to mirror makers in London in the early nineteenth century are still in business, now supplying replacement parts for damaged pieces made from the original molds.

560. Gilded wood, gesso and eglomise, American, probably New York, circa 1800 to 1810. The surge of American patriotism in the early years of the nineteenth century made the American eagle and motto popular decorative motifs. The banner in the eagle's mouth reads *E Pluribus Unum.* The carved detail on this mirror is of exceptionally fine quality. *White House Collection.*

561. Gilded wood, gesso and eglomise, American, probably Massachusetts, circa 1800-1810. Another version of this style, the gesso molded details have fine detail. *White House Collection.*

561

562

563

562 [left] and 563. Gilded wood and gesso, American, bearing the label of John Doggett of Roxbury, Massachusetts, circa 1810. The applied ornaments of an angel and bell-flower swag in the upper panel were molded from gesso and gilded. The label of John Doggett which is on the back of this mirror (563) resembles a girandole mirror. John Doggett (1780-1857) worked in Roxbury and Boston, Massachusetts as a gilder, mirror and picture frame maker, cabinetmaker and merchant between about 1802 and 1846. He worked in partnership with Samuel S. Williams and Samuel Doggett, Junior under the firm John Doggett and Company. From 1825 they also sold carpets. *Courtesy of The Metropolitan Museum of Art, gift of Mrs Russell Sage, 1909 [10.125.377].*

562 (right). Gilded wood and eglomise, American, attributed to Edward Lothrop of Boston, circa 1815. A very similar mirror with Edward Lothrop's label is in the collection of the Henry Ford Museum. The eglomise panel represents the naval engagement between the American ship United States and English ship Macedonian in the war of 1812. *Courtesy of The Metropolitan Museum of Art, gift of Mrs. Russell Sage, 1909 [10.125.378].*

564. Gilded wood and gesso, English, circa 1810. The gesso details made a pleasing arrangement on this frame. The bunch of grapes and leaves were a fairly standard decoration in the first quarter of the nineteenth century. *Courtesy of Alfred Bullard, Inc.*

565. Gilded wood and gesso and eglomise, English or American, circa 1810, 46″ high, 31″ wide. The decorations are each of the finest workmanship; carving, eglomise, and gilding to create a refined mirror. *Courtesy of David Stockwell, Inc.*

564

566

565

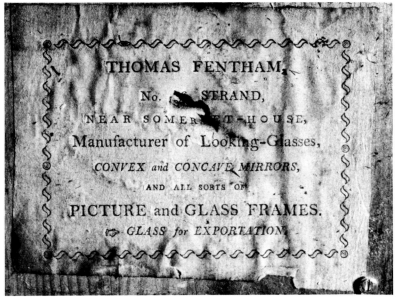

567

566 and 567. Gilded wood and gesso and eglomise, English, bearing the label of Thomas Fentham of London, circa 1805. The Classical decorations of columns, urns and floral swag tied with ribbons are subordinate to the grape vine and geometric pattern in the eglomise panels. As the nineteenth century progressed, larger floral and patterned motifs gradually dominated the decorative scheme. Thomas Fentham's label (567) is particularly interesting because so few English labels were used or are remaining. The address here, 136 Strand, dates this mirror after 1802 when Fentham moved from 52 Strand where he had worked since 1783. *Courtesy of the Colonial Williamsburg Foundation.*

568

569

568. Gilded wood and gesso and eglomise, English or American, circa 1805. The carving is very refined on this mirror and the eglomise glass panel continues the use of geometric details, now with musical instruments as the theme in the center.

569. Gilded wood, gesso, and eglomise, English, circa 1805. The nicely detailed eglomise panel with musical instruments is the dominant feature of this mirror where the frame itself is supportive but contributes little artistic merit.

570. Gilded wood and gesso, and eglomise, English, circa 1800. The late Classical style mirror was also interpreted for overmantles. This example combines all of the features shown in vertical mirrors but in a horizontal design. The eglomise panels here are appropriate replacements.

571. Gilded white pine and eglomise, American, probably Albany or New York City, circa 1805, 81" long, 56" high. This is one of the most ornate and pleasing of the late Classical style American overmantle mirrors. Like mirrors 549 and 550, it was possibly made in Albany or New York City. It was made for New York Governor Joseph C. Yates and his wife who was from Albany. *Courtesy, The Henry Francis du Pont Winterthur Museum.*

570

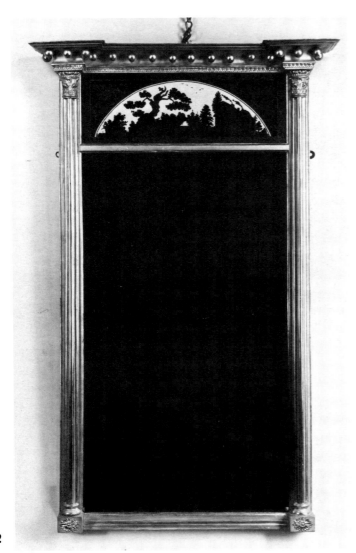

572

573

574

575

576

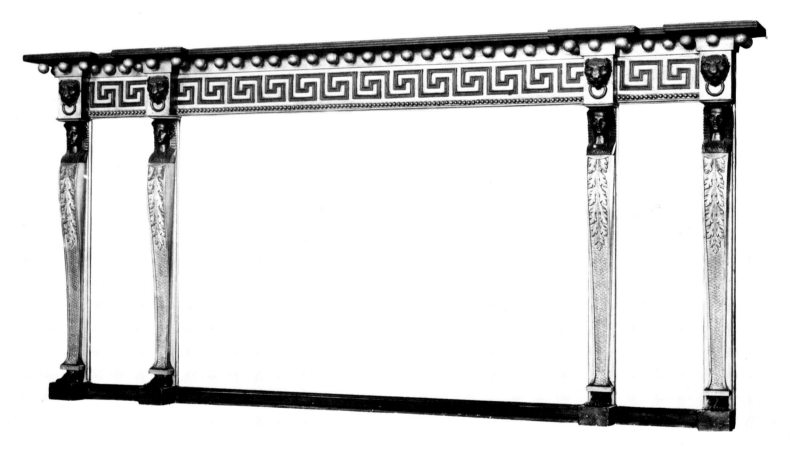

577

572. Gilded wood and gesso, and eglomise, English, circa 1810.

573. Gilded and painted wood and gesso, and eglomise, English, circa 1815. Two important applied decorations are combined on this mirror. The side columns have been modified into Egyptian mummies with black painted heads and feet. Excavations in Egypt were a fascination to the West at this period. Chairs, tables and ornaments on furniture were inspired by the styles found during excavations of ancient tombs.

The caduceus ornaments over each mummy were Greek and Roman herald's staffs as carried by Hermes and Mercury. Today they signify the medical profession.

574. Gilded and painted wood and eglomise, English, circa 1815, 34½" high, 22¾" wide. *Courtesy of Herbert Schiffer Antiques.*

575. Gilded wood and gesso and eglomise, English, circa 1815.

576. Gilded and painted wood and eglomise, English, circa 1815.

577. Gilded and painted wood, English, circa 1815.

The mummies and leopard heads are beautifully executed on these overmantle mirrors which successfully combine Egyptian and Greek (key border motif and oak leaves) motifs. *Courtesy of the Victoria and Albert Museum.*

578

580

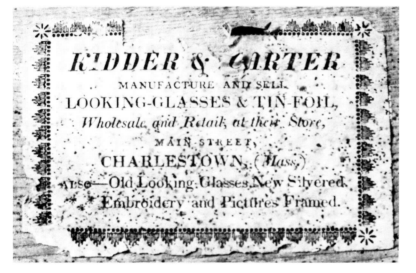

579

578 and 579. Gilded wood and gesso and eglomise, American, bearing the label of Kidder and Carter of Charlestown, Massachusetts, circa 1815. The mirror combines ornate molded gesso and ring turned columns which became ever more common in mid-nineteenth century furniture. This is an early use of molded decorations overlaying the turnings. John Kidder and Joseph Carter were mirror makers and merchants in Charlestown, Massachusetts from about 1811 until 1820. *Courtesy of Herbert Schiffer Antiques.*

580. Painted and gilded wood and eglomise, English, circa 1800. The funny little chinoiserie figure in the center of the eglomise panel is a late carry over from the Chinese Rococo period.

581 and 582. Mahogany and painted wood and eglomise, American, bearing the label of Peter Grinnell & Son of Providence, Rhode Island, circa 1820. This mahogany interpretation of the late Classical mirror has gold details in painted applied decoration and columns and gold eglomise glass.

Peter Grinnell fixed his label to the back and inscribed (or signed?) his name on the frame above the label. Peter Grinnell (1764 to 1836) and William Taylor (d. 1810) worked together as Grinnell and Taylor selling hardware and ship chandlery in Providence, Rhode Island from about 1787 until 1809. From 1809 until about 1828, Peter Grinnell and his son William Taylor Grinnell (1788 to 1835) made frames and mirrors and sold hardware in Providence under the name Peter Grinnell and Son. *Courtesy of Herbert Schiffer Antiques.*

581

583

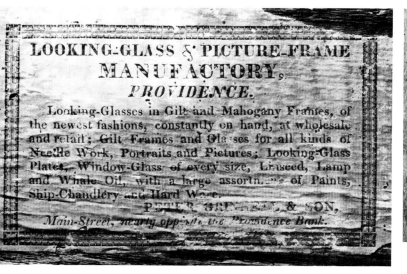

582

584

583 and 584. Painted wood and eglomise, American, bearing the stencilled label of T. Hillier of Pittsburgh, Pennsylvania, circa 1820. The mirror is a typical representative of the late Classical style in America circa 1810-20. The stencilled label (584) brings up ambiguity because a T. Hillier is not listed at No. 7 Fifth Street in Pittsburg directories at this period. However, Thomas A. Hillier, age 47, was listed in the 1850 Pennsylvania census, and T.A. Hillier, looking glass and fancy store is listed on Wood Street in Pittsburg directories between 1837 and 1852. The Philadelphia directories list Thomas A. Hillier variously as coach maker, salesman, agent and grocer from 1828 to 1860. *Courtesy Mr. and Mrs. Samuel Schwartz.*

585

586

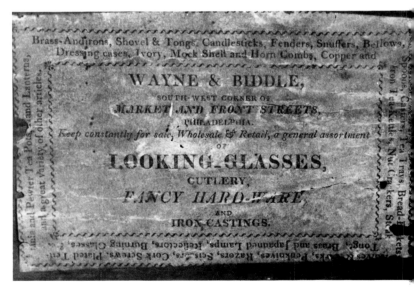

587

585. Mahogany and eglomise, American, probably Philadelphia, circa 1815, 40 9/16" high. The eglomise panel shows a view of George Washington's home Mount Vernon taken from a print published before 1804 when a gallery was added to the roof. See also *552* and *554*. *Courtesy, The Henry Francis du Pont Winterthur Museum.*

586 and 587. Mahogany and eglomise, American, bearing the label of Wayne and Biddle of Philadelphia, circa 1815. This late Classical mirror and 588 were made by Wayne and Biddle at the same time that they were making fret-work mirrors (see 416, 417 and 419). The label on this (*587*) varies from the label *420* only in the last line where "Iron Castings" is included here among their articles for sale. *Courtesy of Herbert Schiffer Antiques.*

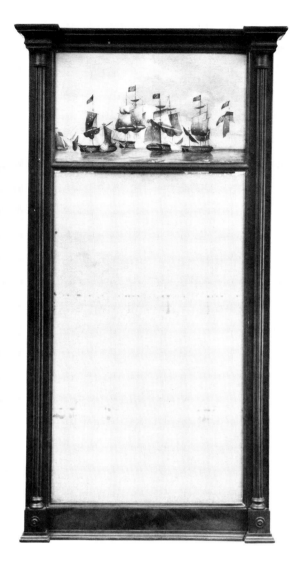

588

589

588. Mahogany and eglomise, American, bearing a label of Wayne and Biddle of Philadelphia, circa 1815, 48 13/16" high, 21½" wide. A navel engagement of the War of 1812 is depicted on this eglomise panel in a frame by Wayne and Biddle. (See also 586). *Courtesy, The Henry Francis du Pont Winterthur Museum.*

589 and 590. Mahogany and eglomise, American, bearing the label of Spencer and Gilman of Hartfort, Connecticut, circa 1815, 38" high, 18 5/8" wide. The naval engagement shown in the eglomise panel is another representation of a battle during the War of 1812 (see 588) with the American eagle and shield. The frame is restrained with only reeded columns as decoration. Stephen Spencer (born circa 1782) and Eli Gilman were brothers-in-law who worked in Hartford, Connecticut from 1808 to about 1827 as mirror makers, importers and merchants. They advertised "30 cases of English looking glasses" [*Connecticut Courant*, April 28, 1818]. Four different labels are known. (See B. Ring, "Checklist...", *"The Magazine Antiques,"* May, 1981.) *Courtesy, The Henry Francis du Pont Winterthur Museum.*

590

591

593

591 and 592. Mahogany and pine, American, bearing the label of W.M. Gaylord of Utica, New York, label dated 1826, 44 1/8" high, 22¾" wide. The carving on these pillars was repeated on drop leaf tables and mantle clock cases of this period. The flat, cut out veneer decoration in the crest is quite unusual, as is the use of oval and round pressed brass bosses above and below the pillars. Wells M. Gaylord made mirrors in Utica, New York at various locations between 1826 and 1845. He worked with Edwin Gaylord between 1839 and 1843. This label (592) is dated January 6, 1826 and lists the factory address at 55 Genesse Street. When the Erie Canal was built, Utica prospered and became a center for manufacturing enterprises. *Courtesy of The Munson-Williams-Proctor Institute, Utica, New York.*

593 and 594. Mahogany and eglomise, American, bearing the label of W.M. Gaylord of Utica, New York, label dated June, 1828, 27" high, 14½" wide. This mirror has the first example of rope carved pillars shown in this book. The popularity of rope carving continued during the mid-nineteenth century on all other forms of furniture. The label (594) of Wells M. Gaylord of Utica, New York (see 591 and 592) is dated June, 1828 when the Gaylord factory continued at 55 Genessee Street. *Courtesy of The Munson-Williams-Proctor Institute, Utica, New York.*

595 and 596. Mahogany and printed paper, English, circa 1820, 46½" high, 26" wide. The carved frame includes ring turned and rope carved pillars and drops which are acorn shaped. The upper glass panel is particularly interesting because it encloses

592

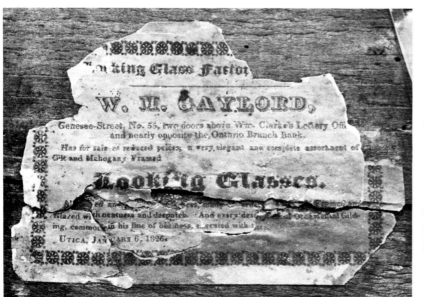

594

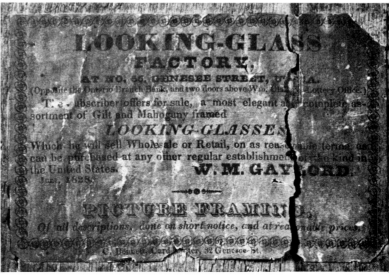

595

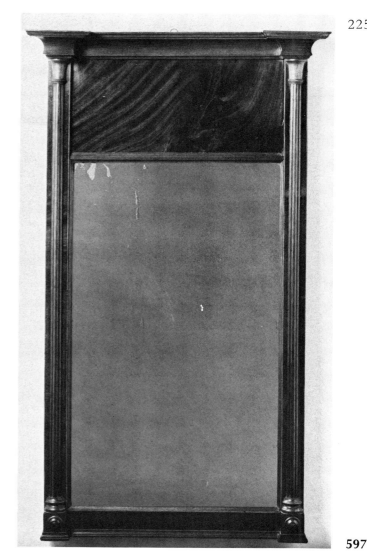

597

a tinted print (596) of "RELIGION" and "BRITTAN-NIA" lamenting the death of her late KING GEORGE the third, born June 4th, 1738, ascended the throne 25th of October 1760, (died) 9th of January, 1820 in the 60th year of his reign. Published and Painted by W.B. Walker, 4 Fox and Knot Court, Cow Lane, London, "entitled "IN MEMORY OF OUR GOOD OLD KING." *Courtesy of the Anglo-American Art Museum, Memorial Tower, Louisiana State Museum, Baton Rouge, Louisiana.*

597 and 598. Mahogany and pine, American, bearing the label of Henry M. Fisher of Baltimore, Maryland, label dated July 14th, 1812, 38" high, 21" wide. This mirror exhibits the brilliant choice of highly figured mahogany veneer which characterizes Baltimore cabinet work of this period. The otherwise restrained frame shows off a lively wood

grain in the upper panel. The label (598) is dated July 14th, 1812, and locates the store at 38 South Street. Henry M. Fisher worked as a carver, gilder, painter and mirror merchant in Baltimore at various addresses between 1802 and 1829. *Courtesy, The Henry Francis du Pont Winterthur Museum.*

598

596

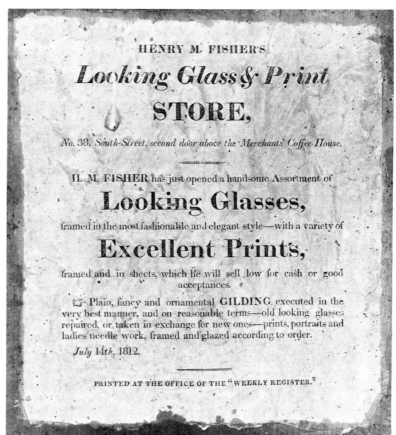

HENRY M. FISHER'S

Looking Glass & Print

STORE,

No. 38, South-Street, second door above the Merchants' Coffee-House.

H. M. FISHER has just opened a handsome Assortment of

Looking Glasses,

framed in the most fashionable and elegant style—with a variety of

Excellent Prints,

framed and in sheets, which he will sell low for cash or good acceptances.

☞ Plain, fancy and ornamental GILDING executed in the very best manner, and on reasonable terms—old looking glasses repaired, or taken in exchange for new ones—prints, portraits and ladies needle work, framed and glazed according to order.

July 14th, 1812.

PRINTED AT THE OFFICE OF THE "WEEKLY REGISTER."

599

600

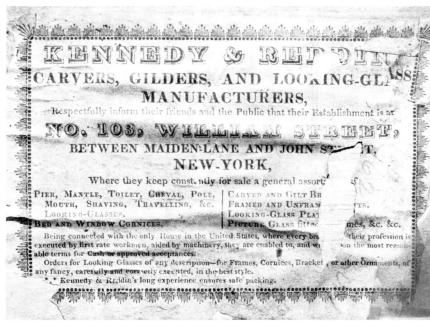

601

599. Painted and gilded wood, English or European, circa 1800 to 1825, 88¾" high, 40½" wide. Microanalysis has determined that this mirror was made from European wood, yet it was most definitely made for the American market. The American eagle crest is almost gaudy in its boldness and painted detail. European goods with American patriotic motifs found a ready market in America during the first quarter of the nineteenth century. Other examples of the trade west include Liverpool-made pottery jugs with American patriotic scenes. *Courtesy, The Henry Francis du Pont Winterthur Museum.*

600 and 601. Mahogany veneer and gilded wood, American, bearing the label of Kennedy and Reddin of New York, circa 1827. The last remnants of Classical style seem to be the overhanging cornice and side columns, which now have become a serpentine outline and thick, unornamented smooth columns respectively. The applied decoration is heavy and arranged more than designed. The label (*601*) fixes this mirror as the work of a partnership between Thomas Kennedy and an as yet unidentified Mr. Reddin in New York in 1827. Thomas Kennedy was a mirror and frame maker working between about 1827 and 1865. He had

many locations during that period, but was at 103 William Street only in 1827 with Mr. Reddin.

The label is particularly interesting because it lists a variety of mirror types for sale. The last line reading "Kennedy & Reddin's long experience ensures safe packing" suggests the partnership had been together for a number of years by 1827. *Courtesy of Herbert Schiffer Antiques.*

603

Convex

602

Convex, round mirrors in the early nineteenth century evolved from the Renaissance tradition (see 11 and 12) and French prototypes of the late eighteenth century. The French glass factories produced convex glass inexpensively enough to make it popular, and when supplies were cut off to England after the French Revolution, the English developed their own processes and supplies. Ince and Mayhew included designs for convex mirrors in their *Universal System* in 1759-1762, but the fashion for this style did not prevail in England or America until about 1800. Many of the Classical details associated with other mirror styles were also incorporated on convex mirror frames.

602. Gilded wood, English or Continental, circa 1790.

The standard frame for convex mirrors of this period has a round, continuously convex molding with contrasting dark ribbed fillet next to the glass, and ornaments in an upper and a lower crest. Because so many mirrors of this type also support candle arms, the style is often known as girandole, and that term has come to apply to convex mirrors of this period both with and without candle arms. For some reason as yet obscure to the author, sea creatures, both real and imaginery, are often used as decorations on these mirrors. Mirror *602* has a pair of sea horses flanking a deer in the upper crest, and a shell motif reminiscent of Rococo design in the lower crest. *Courtesy of Alfred Bullard, Inc.*

603. Gilded wood and gesso, English or Continental, circa 1790.

The prolific amount of decoration on this mirror renders a shimmering effect to the surface. The sea horse appears again (see 602) above rocky ledge and tightly scrolled leaves. The frame is ornamented with a honey-comb pattern in relief. The four candle arms gracefully emerge from the leaves of the lower crest and support brass candle cups and crystal bobeches with pendant prisms. The effect is dazzeling. *Courtesy of Alfred Bullard, Inc.*

604

605

606

604. Gilded wood and gesso, English or American, circa 1800.

A pair of intertwined dolphins on a rocky mount ornament this crest over a relatively restrained frame and pair of candle arms. American white pine was used by frame makers at this period on both sides of the Atlantic Ocean. Until a quantity of labeled or otherwise documented convex mirrors are found to be analyzed and compared, the precise country of origin remains obscure. *Courtesy of Herbert Schiffer Antiques.*

605. Gilded wood and gesso, English, circa 1800.

Black balls against a gilded frame and the lattice work in the concave molding are unusual features of this mirror. Probably of English origin, it is assumed to have been made for the patriotic American market, since the eagle is so dominant as to pull the crest out of proportion. These graceful candle arms support fine cut glass candle cups, bobeches and pendant prisms which are most likely English or Irish crystal. *Courtesy of Alfred Bullard, Inc.*

606. Gilded wood, English or American, circa 1800.

Gilded, applied, ball ornaments were common on convex mirrors as they were on the tabernacle mirrors being made at this same period. This is a good example of a typical form. *Courtesy of Alfred Bullard, Inc.*

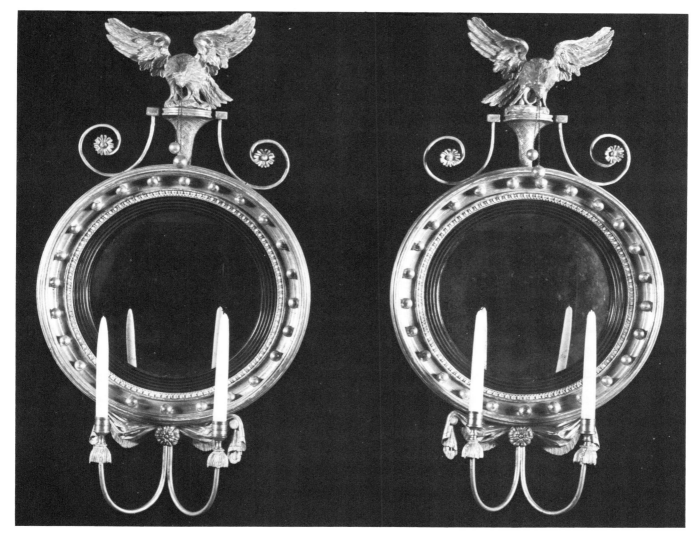

607

607. Gilded wood, gesso and wire, English, circa 1815. The eagles hold in their beaks chains with gilded balls at each end said to represent chained shot which was used to demast rival ships during the Napoleonic and 1812 Wars. *Courtesy of M. Turpin Antiques.*

608. Gilded wood and gesso, American, circa 1800, 44½″ high. The finely carved gesso borders and light crest decoration give a graceful appearance to this convex mirror. An early Philadelphia newspaper page, thought to be original backing material, was found behind the glass, and suggest an American origin. This form of mirror was made in Russia, Italy, and England besides America with a wide variety of sizes.

608

609

610

611

609. Gilded and painted wood and wire, English, circa 1800, 51" high, 23½" wide. The graceful lyre in the crest and delicate prisms on the candle cups give a genteel appearance to this mirror. The supports for the candle cups terminate in black painted animal heads.

610. Gilded wood and gesso, English, circa 1800, 100" high, 42" wide. This is an enormous mirror of fine quality with carved leaves of the best Rococo style. Thomas Sheraton mentions in his *Cabinet Directory* of 1803 that convex mirrors had become unusually popular for their "Agreeable" effect by reflecting the perspective of rooms and their convenience as a light source. The quality, condition and originality of the gilding makes a significant difference in value among all of the gilded mirror types. *Courtesy of the Victoria and Albert Museum.*

611. Gilded and painted wood and gesso, English, circa 1800, 59" high, 36" wide. The Classical motifs of bows and wreaths are applied in the fashion of French ormolu mounts to this dark painted concave molding. The carving in both crests is of superb quality.

612. Gilded and painted wood, English or American, circa 1800. *Courtesy of Alfred Bullard, Inc.*

613. Gilded wood and wire, English, circa 1800, 38" high, 26½" wide. *Courtesy of Alfred Bullard, Inc.*

614. Gilded and painted pine, English, circa 1800, 36" high, 22" wide. There is an old story told that convex mirror frames were painted black in America when George Washington died (1799) and in England when Admiral Nelson died (1805), yet a reliable reference for this has not come to the attention of the author. *Courtesy of Herbert Schiffer Antiques.*

615. Gilded and painted wood and gesso, English or American, circa 1820, 33½" high, 18¼" wide. The pair of serpents in the lower crest continue the marine theme of decorations on these convex mirrors. *Courtesy of Herbert Schiffer Antiques.*

612

613

614

615

616

616. Gilded and painted wood, English or American, circa 1800. *Courtesy of Alfred Bullard, Inc.*

617. Gilded and painted wood, English or American, circa 1820, 61" high. The frame is a concave molding rather than convex and the decoration has little direct relation with the Classical motifs seen on earlier mirrors of this style. *Courtesy of H. and R. Sandor, Inc.*

618

617

618. Barometer, mahogany, English, circa 1800, 39½" high. This five-dial, banjo-shaped barometer is representative of the group of barometers which often included convex mirrors in the first quarter of the nineteenth century. The top dial is a dry-damp guage, then a thermometer with calibrations in Farenheit degrees and comments (for the English climate) freezing, temperate, summer heat and blood heat. Below the convex mirror is the barometer dial, and at the bottom a level. *Courtesy of Herbert Schiffer Antiques.*

REVIVALS AND INNOVATIONS

619. Mahogany and gilded brass, French, early nineteenth century. The French Empire style was a distinct and separate outgrowth of the European Classical style identified by mahogany veneer typically overlaid with gilded brass (ormolu) mounts. It is particular in its overall box-like, rectangular forms. The ormolu decorations at the earliest phase of the Empire style are crisply cast with precise details carefully represented. Figured moldings and individual decorations are combined in symmetrical and restrained arrangements. This cheval glass (619) is a fine quality Empire design. The urn finials, candle arms, post supports and foot scrolls have the best quality castings and the mirror border and ormolu mounts in the base are restrained and elegant. *Courtesy of The Metropolitan Museum of Art, Rogers Fund, 1920.*

620 and 621. Mahogany veneer on pine, American, bearing the label of S. Nolen of Philadelphia, circa 1830 to 1850, 22¼" high, 23¼" wide, 7 13/16"

deep. This dressing mirror is an American interpretation of the French Empire style. The overall design is box-like rectangular in mahogany veneer with French style cast brass feet, escutcheon and knobs. The rest of the design is an approximation of French Empire in painted spandrels and turned supporting posts. The stringing inlay is a carry over from the American Hepplewhite style. It is a pleasing mirror with these composit details. Spencer Nolen made mirrors and ornamental painting in Boston between 1804 and 1816 and Philadelphia thereafter until 1850. He worked with Aaron Willard (1806) and Samuel Curtis (1806-1822) making clocks and painting signs in Boston. When he moved to Philadelphia, Spencer Nolen continued a branch of Nolen & Curtis making clocks until 1822 when he is listed independently as a looking glass manufacturer. From 1828 until 1849 he worked at 78 Chestnut Street, Philadelphia as this label (621) records. *Courtesy, The Henry Francis du Pont Winterthur Museum.*

620

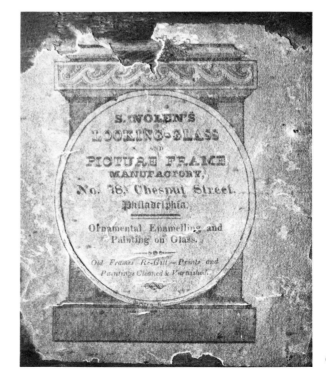

622

623

624

625

625. Mahogany, marble, gilded wood and brass, American, attributed to Anthony Quervelle of Philadelphia, circa 1835. The mirrored back panel of this pier table reflects the beautifully carved and gilded front legs which are the dominant decorative features of this Empire style design. Brass inlay also ornaments the skirt, wood below the glass and shelf edge. Anthony Quervelle was a very successful French cabinetmaker who worked in the Empire style in Philadelphia between 1835 and 1849. The elaborate carving is representative of his work. *Courtesy of David Stockwell, Inc.*

626. Mahogany, marble and gilding, American, attributed to Anthony Quervelle of Philadelphia, circa 1840. This pier table again has a mirrored glass fitted to the back panel, but the design has evolved to include a scalloped shape at the shelf, exclusively carved ornaments including grapes, and massive animal's paw feet, in fact two pairs of feet. Empire restraint is giving way in this design to a revival of Rococo exuberance. *Courtesy of David Stockwell, Inc.*

626

622. Maple veneer, American, circa 1825, box closed 12 1/8" wide, 4¼" high, 10 3/16" deep. The hinged lid protects the folding mirror for travel. This design is probably an innovation on the part of a skilled rural, New England cabinetmaker yet it is in the tradition of the Empire style's use of figured wood as decoration. *Courtesy, The Henry Francis du Pont Winterthur Museum.*

623. Mahogany, marble and brass, American, circa 1820. The upright back panel of this American Empire pier table is mirrored glass framed by the marble top and mahogany back legs and shelf. Like most furniture of this style, the wood was carefully chosen for its brilliant figure. The column bases and caps and the escutcheon and bosses are finely cast brass of probable European origin imported to America. *Courtesy of Samuel T. Freeman Company, Auctioneers.*

624. Mahogany and brass, American, New York, circa 1830, pier table 42" wide, 37" high, 16" deep; secretary 36" wide, 70½" high, 18½" deep. The upright back panels of the pier table and secretary and the doors of the secretary are mirrored glass. Complete sets of matching furniture were designed in the Empire style to coordinate room settings. In this case, all of the decorations were marked French brass castings of a very high quality incorporated into the furniture made in New York for members of the Delafield family. *Courtesy of David Stockwell, Inc.*

627. Wood, Chinese, circa 1840. This Chinese-made dressing stand follows a design used in England from about 1760. It was probably special ordered originally, and later copied for trade with America. The mirror slides up from the back and pivots within the outer molding. *Courtesy of The Museum of American China Trade.*

628. Walnut and gilt, American, attributed to Richard Upjohn, Circa 1860, 67" high, 45" wide. This mirror in the American Gothic style was designed for the library of the Robert Kelly house of New York ensuit with the cornice and book cases which are not shown. The Gothic style in America was popular principally in New York state between 1830 and 1840 after Andrew Jackson Downing published designs in this style in *Cottage Residences* and *Architecture of Country Homes.* Richard Upjohn lived between 1802 and 1878 and worked as a leading architect in New York City. Among his famous interiors are those for Trinity Church on lower Broadway and the Church of the Ascension in New York. *Courtesy of Munson-Williams-Proctor Institute, Utica, New York.*

629. Gilded wood, French, circa 1820, 85" high, 35½" wide. Rectangular mirrors with gilded molded frames otherwise not ornamented appear vertically as well as horizontally in pictures of room interiors thoughout Europe and America in the early nineteenth century. Some were used in the most lavish interiors. This mirror was purchased in France by Stephen Girard, a wealthy merchant, for his home in Philadelphia between 1810 and 1820. *Courtesy, The Stephen Girad Collection, Girard College, Philadelphia.*

627

628

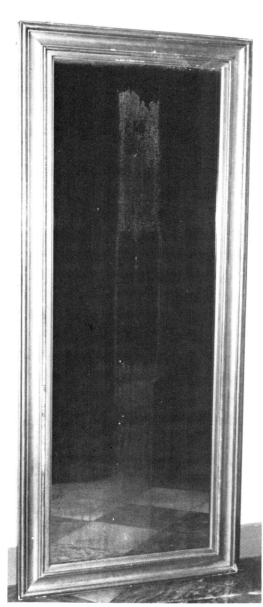

629

630

630. Maple and pine, American, attributed to Eneas P. John of Londonderry township, Chester County, Pennsylvania, circa 1850, 17 3/8" high, 14 3/8" wide. There are countless mirrors with molded frames similar to this in America for it was a rural style popular for a hundred years from 1770. This one has a history linking it with cabinetmaker Eneas P. John who worked for local orders west of Philadelphia. He is mentioned in an 1849 tax assessment and an inventory in 1862 lists no cabinet-making tools. *Courtesy, Mr. Norman S. Pusey.*

631. Rosewood veneer on pine and gilding, American, Pennsylvania, circa 1840, 48½" high, 28¼" wide. Thousands of mirrors of this type were made when manufacturing enabled mass-production of moldings. This type is often called an "ogee" mirror because the molding is "ogee" shaped (serpentined). One owned by the Philadelphia Museum of Art is stencilled "H. & J. Meyers, Mount Joy" (Pennsylvania). (See *"The Magazine Antiques,"* April, 1926.) *Courtesy of Herbert Schiffer Antiques.*

632 and 633. Gilded wood, American, bearing the label of Paul Mondelly of Boston, circa 1850. This simply molded frame is gilded and supported by a brass loop screwed to the top of the frame. This style was popular throughout the mid-nineteenth century for it blended with the various designs of interior style which were all being made simultaneously. The label (633) indicates that Paul Mondelly worked at 73 Cornhill, Boston making and selling looking glasses, prints and supplies. *Courtesy of Herbert Schiffer Antiques.*

631

632

PAUL MONDELLY,

No. 73, CORN......

KEEPS constantly for sale, at the most reduced prices, a complete assortment of LOOKING-GLASSES....PRINTS....PICTURE GLASSES... SPY GLASSES....THERMOMETERS....Drawing PAPER....PAINTS....PENCILS, &c. &c.... with all kinds of FRAMES in his line. ALSO,

LADIES' NEEDLE WORK handsomely framed in the most modern style, and at the shortest notice—OLD FRAMES new gilded—OLD GLASSES new silver'd.

633

634

636

EARPS & Co.
South West
CORNER OF MARKET AND FIFTH ST.
PHILADELPHIA,
Importers of Hardware and Cutlery;
AND MANUFACTURERS OF
GILT AND MAHOGANY FRAME
LOOKING GLASSES;
Brass Andirons, Tongs and Shovels, Fenders,
&c. &c.
WHOLESALE AND RETAIL.
Latourette, Printer, N. W. corner of 2d & Race sts.

635

637

638

639

634 and 635. Gilded wood, American, bearing the label of Earps & Co. of Philadelphia, circa 1825. This is a delicate version of the previous (632) plain molded frame. The edges of the frame are marked with a saw-tooth line and a spiral twist turned molding is added. The acorn drops above and the ball drops below are not ordinary. The label (635) identifies Earps & CO. at an address they occupied in 1825, and it states that they made gilt and mahogany mirrors and brass fireplace equipment. George, Thomas and Robert Earp and Robert's widow were collectively hardware merchants and mirror manufacturers in Philadelphia between 1813 and 1825. They used the names Earps & Co. (1814-17 and 1825) Earp & Baxter (1818) and Earp & Brothers (1818-24) at various locations. *Courtesy of the Philadelphia Museum of Art, given by Mrs. Alfred Coxe Prime.*

636. Gilded wood and gesso, American, bearing the label of Spencer Nolen of Philadelphia, circa 1830, 36 7/8" high, 24 5/8" wide. These thick turned columns in the frame with applied gesso decorations were popular in the late Empire period about 1830 to 1850. They relate the mirror to furniture with boldly turned legs. Spencer Nolen's label (see also *621*) on the mirror indicates that this style was made in Philadelphia. *Courtesy, The Henry Francis du Pont Winterthur Museum.*

637. Gilded wood, American, circa 1820. The anguished swans' heads completely dominate the design of this mirror and relate it to other carved furniture of this period with swan's necks as ornaments. The eagle, although proportionally large and well carved, serves only to fill the space between the swans' heads, but adds nothing to improve the design. *Courtesy of Joan Bogart Antiques.*

638 and 639. Gilded wood and gesso, American, bearing the label of Isaac L. Platt of New York, circa 1830, 71" high, 38" wide. This is quite a large example of late Empire style mirrors and includes, besides the convex molded frame, applied decorations and corner ornaments. It was made by Isaac L. Platt for the Randall Mansion in Cortland, New York. Isaac Platt (1793 to 1875) made mirrors and frames and sold prints in New York City between 1815 and about 1860 at various locations. He also helped found the Chemical Bank and the Pennsylvania Coal Company. The label (639) on this mirror gives the address 178 Broadway where he is listed between 1825 and 1837. *Courtesy of the Munson-Williams-Procter Institute, Utica, New York.*

640

642

641

640 and 641. Mahogany and mahogany veneer with marble, American, bearing the label of John Needles of Baltimore, Maryland, circa 1840, 79 3/8" high, 46" wide, 23 3/8" deep. The frame and supports of this mirror are carved in the American Rococo style which evolved between 1840 and 1870 throughout the country. This mirror is joined to the white marble top of the mahogany veneered chest to create a dressing bureau. John Needles (1786-1878) had a cabinet shop in Baltimore between 1808 and 1850. He worked at 10 Hanover Street between 1808 and 1812, and at 54 Hanover Street between 1812 and 1850 when his business was taken over by Thomas Godey, a cabinetmaker whose shop was at 54½ Hanover Street. The label (*641*) is quite faded, but reads John Needles/Manufacturer/of/Cabinet Furniture,/10 Hanover Street/Baltimore. The two sofas below this inscription are designed in the Rococo and Gothic revival styles. *Courtesy, The Henry Francis du Pont Winterthur Museum.*

642. Oak, American, by John Henry Belter, circa 1850. The Rococo revival style of the nineteenth century in America is no better exemplified than by the work of John Henry Belter. This hall stand with mirror is dominated by c-scrolls, curving form and carved detail to the utter suppression of the function of the piece. John Henry Belter was a German immigrant who worked in New York City between 1844 and 1861. He was granted several patents including one for his original laminated construction which enabled him to curve sheets of wood on two planes. *Courtesy of Joan Bogart Antiques.*

643

643 and 644. These two views of a room setting at the Metropolitan Museum of Art (gallery 12) display furniture made by John Henry Belter and Rococo revival style mirrors. The rectangular pier mirror in *643* and oval overmantle mirror in *644* are both gilded wood and gesso of particularly fine detail dating about 1850. Also in 644 is a display shelf over chest made by John Henry Belter with a mirrored back panel (see *642*). *Courtesy of The Metropolitan Museum of Art.*

644

645

645 and 646. Gilded wood, American, made by Thomas J. Natt of Philadelphia in 1846. The over-mantle mirror (*645*) and pier mirror (*646*) were billed by Thomas J. Natt of Philadelphia in 1846 to the Logan family of "Loudon" in the Germantown section of Philadelphia for whom they were made. They are designed in the Rococo revival style of Louis XV designs and are particularly nicely carved. The crests of the two mirrors match, although of different proportions.

Thomas J. Natt (circa 1805-1859) succeeded his father Thomas Natt (see 527 and 528) as a looking glass maker and merchant between about 1833 and 1855. *Photography by Courtlandt V.D. Hubbard.*

646

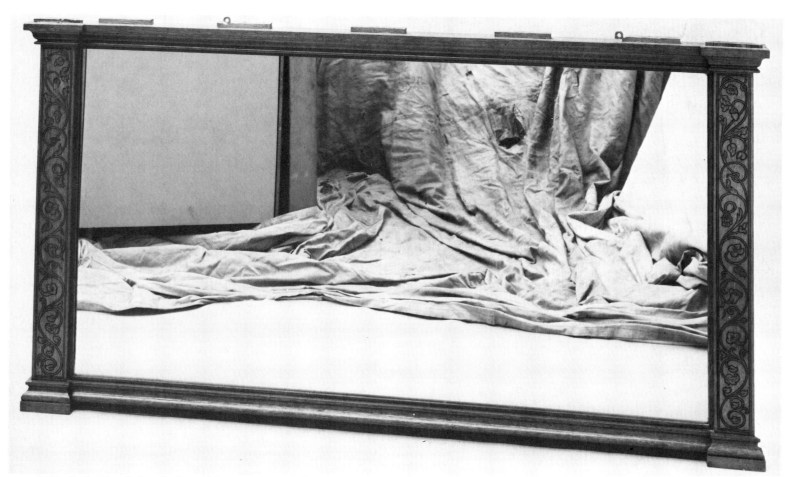

647

647. Oak, English, designed by Antony Salvin, in 1838. The Elizabethan-revival style of the nineteenth century is clearly demonstrated in these next three examples. The incised decoration (647) is in a tight meandering style inspired by Elizabethan carving (see 15).

Anthony Salvin (1799-1881) made this mirror and 649 for "Manhead," South Devon which he designed throughout after Elizabethan prototypes. *Courtesy of the Victoria and Albert Museum.*

649

648. Oak, English, circa 1840. The use of oak, precise carving and twisted columns demonstrate a revival of the Elizabethan style. *Courtesy of the Victoria and Albert Museum.*

649. Oak, English, designed by Anthony Salvin in 1838. See 647. *Courtesy of the Victoria and Albert Museum.*

648

650

653. Gilded cardboard, German, circa 1900, 2" high, 1½" wide. This little mirror is a Christmas tree ornament probably made in the Dresden-Leipzig area of Germany from embossed cardboard. *Courtesy of Mrs. Herbert F. Schiffer.*

654. Animal horn, teeth, and wood, Austrian or German, circa 1865. Animal horn furniture was first popularized by the romance of the American western frontier. Horns of Texas longhorn cattle were used to make hall stands, chairs, and mirror frames which satisfied a public searching for associations with the rustic life of hunters and cattlemen. Austrian manufacturers also produced horn furniture which sold at Tiffany and Company in New York City intended to furnish Baronial mansions, hunting lodges, libraries, and gentlemen's billiar rooms. This mirror (654) has elk horns and boars' teeth. *Courtesy of the Victoria and Albert Museum.*

655. Enamel on copper and embroidered silk, Chinese, circa 1840, 14" long, 6½" wide. Chinese craftsmen made many hand mirrors from wood, ivory, jade, etc. for sale to foreign merchants. The frame of this hand mirror is Peking enamel in polychrome floral and butterfly design over copper. The back has a silk panel decorated with embroidered goldfish and flowers. *Courtesy of Herbert Schiffer Antiques.*

650. Gilded veneered and inlaid wood, marble, and brass, English, by Jackson and Graham, in 1855, about 156" high. This Renaissance revival mirror and cabinet was made as a showpiece to demonstrate their technical skill by the firm Jackson and Graham and exhibited at the Paris Exhibition of 1855. It draws design inspiration chiefly from Renaissance case pieces (see *20*) but a close study of the elements includes Classical motifs and Rococo architectural details. *Courtesy of the Victoria and Albert Museum.*

651. Gilded, veneered and inlaid wood and marble, English, made by Gillow and Company, circa 1867. Here is another Renaissance revival mirror and cabinet made to exhibit technical skill. Cost and time were not factors of the design or its execution. The English firm Gillow and Company exhibited this at the 1867 Paris Exhibition. The front panels of the cabinet are decorated with inlaid allegorical figures of the arts labeled "Architectura" and "Pintura". *Courtesy of the Victoria and Albert Museum.*

652. Mahogany and contrasting wood inlay, Dutch, mid-nineteenth century. The mirrored door of this Dutch corner cupboard is shaped and decorated to resemble seventeenth century marquetry. (See *27, 31* and *32*).

651

652

653

654

655

656

656. Painted oak and pine, English, circa 1880. A fascination with Egyptian art developed in Europe as a consequence of the English Royal Geographical Society expedition to discover the source of the White Nile River in 1858 and the missionary work of David Livingstone (1813-1873) in Africa. By 1848, the English were helping to organize the Egyptian government and build a railroad and dams across the Nile River. Back home, the exotic culture was exhibited in furniture, this wardrobe (656) a particularly impressive example. (See also 573 to 577).

657. Carved and painted wood, English, 1877. A mirror is incorporated in the right wall of this, the ante-room at "The Grove," Harbourne, Birmingham which was designed in the Neo-Grec style by J.H. Chamberlain (1831 to 1883) in 1877. The carving was done by Mr. Barfield of Leicester. *Courtesy of the Victoria and Albert Museum.*

657

658, 659, and 660. Carved wood, American, by Keller, Sturm and Ehman. Chicago, Illinois, 1885.
These mirrors appeared in the 1885 catalog of pier mirrors, overmantle mirrors, hat trees and cabinets made by Keller, Sturm and Ehman of Elizabeth and Fulton Streets in Chicago. They exemplify the machine made mass produced mirrors which were available to the general public through trade magazines, national showrooms, and mail order catalogs. The development of mechanical manufacturing brought the price per item down to a level where the common man could buy furniture his father only dreamed about.

658

659

660

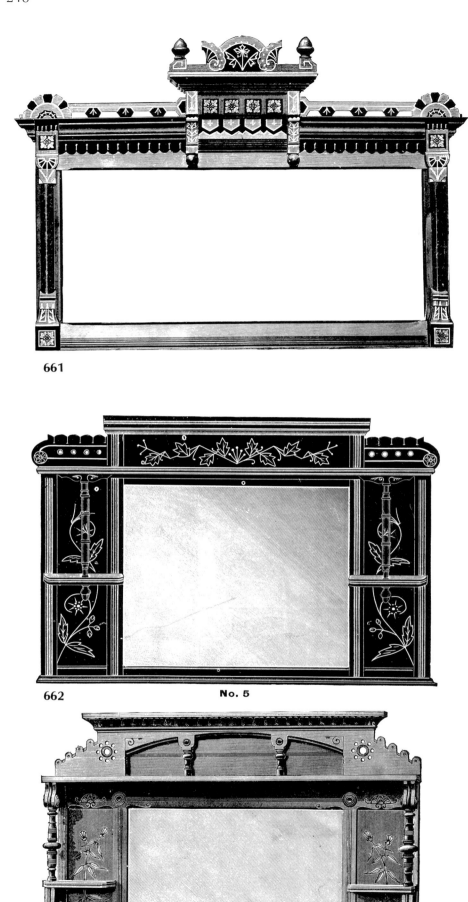

661

662 No. 5

663

664

665

661 through 671. Carved wood, American, by Keller, Sturm and Ehman. Chicago, Illinois, 1885. Like those on the preceeding page, these mirrors and hall stands with mirrored back panels (669, 670, and 671) appeared in the 1885 catalog of Keller, Sturm and Ehman.

666

667

668

669

670

671

672

673

674

675

676

672. Walnut or quartered oak, American, by Bardwell, Anderson and Company, Boston, Massachusetts, 1884, 48" wide, 18" deep, 88" high. This desk is an American adaptation of the Renaissance style mass produced for the general public.

673 to 676. Hardwood in imitation of walnut, mahogany, or oak, American, by M. Samuels and Company, New York, New York, 1890. These are four folding beds with mirrored glass panels mounted to the fronts (under when the bed is folded down). They came with standard 18" x 40" German beveled plate glass or 20" x 40" German glass at $3.00 extra, 18" x 40" French glass at $2.00 extra, or 20" x 48" French glass at $5.00 extra.

677. Mahoganized cherry, walnut or antique oak, American, by Tyler Desk Company, St. Louis, Missouri, 1889, 35" wide, 22" deep. The beveled French glass in the top is 12 inches in diameter. This desk sold for $55.00 in 1889.

678. Chiffonier, American, by Cron, Kills and Company, Piqua, Ohio, 1890, glass panels 8" square.

679 and 680. Dressers, American, by Kent Manufacturing Company, Grand Rapids, Michigan, 1887.

677

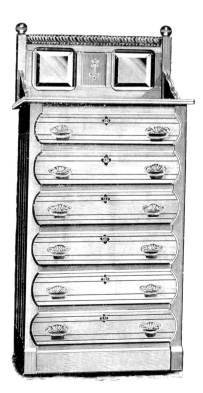

678

679

680

681

682

683

681. Oak sideboard, American, by C. and A. Kreimer Company, Cincinnati, Ohio, 1890, 79" high, 48" wide, 24" deep, glass 10" high x 36" wide and 18" high x 36" wide.

682. Oak sideboard, American, by C. and A. Kreimer Company, Cincinnati, Ohio, 1890, 84" high, 54" wide, 24" deep, glass 20" high, 42" wide.

683. Wardrobe, American, C. and A. Kreimer Company, Cincinnati, Ohio, 1890, 104" high, 58" wide, glass 16" wide, 54" high.

684. Washstand, American, C. and A. Kreimer Company, Cincinnati, Ohio, 1890, 39" wide, 19" deep, glass 22" high, 26" wide.

684

685

685. Mahogany, American, circa 1880 to 1900, 61¾" high, 42" long, 23" deep. The Rococo revival in the mid-nineteenth century in America gave a basis from which the Art Nouveau style emerged. This tall dressing mirror has elements of the French Louis XV Rococo style in the scrolled feet, brass pulls and crest carving, but the proportions and shallow carving are derived from the Art Nouveau designs. *Courtesy of the Newark Museum.*

686. Veneered wood, English, made by Neals Maccasad, circa 1930. The Art Deco style generated from the late nineteenth century designs as a reaction to Rococo. This cheval dressing glass is a new interpretation of the Classical style a hundred years previously (see 543 and 544). *Courtesy of the Victoria and Albert Museum.*

687. Tulip wood veneer and ivory, German, designed by Bruno Paul and made by Herrmann Gerson of Berlin, circa 1924, table 53½" high, 48 5/8" wide. This dressing table with attached mirror and stool are refined examples of the Art Deco style which was advanced in Germany in the first quarter of the twentieth century. *Courtesy of The Metropolitan Museum of Art, gift of Ralph and Kester, Weilding, 1976, in memory of Daly Weilding [1976. 288. 1ab,.2]*

688. Silver, American, circa 1910. Art Nouveau hand mirrors became popular with complete dressing sets in the early decades of the twentieth century. *Courtesy of Mrs. Madeleine Gimble.*

686

687

688

BIBLIOGRAPHY

Bates, Elizabeth Bidwell and Fairbanks, Jonathan L. *American Furniture, 1620 to the Present.* New York: Richard Marek Publishers, 1981.

Bjerkoe, Ethel Hall. *The Cabinetmakers of America.* Exton: Schiffer Publishing Limited, 1978.

Carrington, C.E. and Jackson, J. Hampden. *A History of England.* Cambridge: At the University Press, 1954.

19th-Century America, Furniture and Other Decorative Arts. New York: The Metropolitan Museum of Art, 1970.

Cescinsky, Herbert. *English Furniture from Gothic to Sheraton.* New York: Dover Publications, Inc., 1968.

Chippendale, Thomas. *The Gentleman and Cabinet-Maker's Director.* New York: Dover Publications, Inc., 1966; third edition.

Comstock, Helen. *The Looking Glass in America, 1700-1825.* New York: Viking Press, 1968.

Comstock, Helen. *American Furniture,* Seventeenth, Eighteenth, and Nineteenth Century Styles. New York: Viking Press, 1962.

Constantino, Ruth. *How to Know French Antiques.* London: Peter Owen, 1963.

Dubrow, Eileen and Richard. *American Furniture of the 19th Century.* Exton: Schiffer Publishing Limited, 1983.

Edwards, Ralph, and Ramsey, L.G.G. Editors. *The Connoisseur's Complete Period Guides to the Houses, Decoration, Furnishing and Chattels of the Classic Periods.* New York: Bonanza Books, 1968.

Edwards, Ralph. *The Shorter Dictionary of English Furniture from the Middle Ages to the Late Georgian Period.* London: Country Life Limited, 1964.

Fleming, John. *Robert Adam and his Circle in Edinburgh and Rome.* London: John Murray, 1962.

Gorlich, G.G. Editor. *Il Mobile Veneziano del 1700.* Milano, 1958.

Hartlaub, G.F. *Zauber des Spiegels; Geschichte und Bedeutung des Spiegels in der Kunst.* Munchen: R. Piper and Company, 1951.

Hayward, Helena and Kirkham, Pat. *William and John Linnell, Eighteenth Century London Furniture Makers.* London: Studio Vista and Christie's, 1980.

Haywood, Helena. *The Elliotts of Philadelphia.* M.A. thesis for University of Delaware, Winterthur Program.

Heal, Sir Ambrose. *The London Furniture Makers, from the Restoration to the Victorian era, 1660-1840.* London: B.T. Batsford, Limited, 1953.

Hepplewhite, George. *The Cabinetmaker and Upholsterer's Guide.* New York: Dover Publications, Inc., 1969. 3rd edition of 1794.

Hinckley, F. Lewis. *A Directory of Antique French Furniture, 1735-1800.* New York: Crown Publishers, Inc., 1967.

Hogarth, William. *The Works of William Hogarth in a series of Engravings: with Descriptions, and a comment on their Moral Tendency, by the Reverand John Trusler, to which are added Anecdotes of the author and his works, by J. Hogarth and J. Nichols.* London: Jones and Company, 1833.

Hope, Thomas. *Household Furniture and Interior Decoration, Classic Style Book of the Regency Period.* New York: Dover Publications, Inc., 1971.

Ince & Mayhew. *The Universal System of Household Furniture.* London: Alec Tiranti, 1960.

Kisa, Anton. *Glas im Altertume,* 2 volumes. Leipzig: Karl W. Hiersemann, 1908.

Montgomery, Charles F. *American Furniture The Federal Period in the Henry Francis duPont Winterthur Museum.* New York: The Viking Press, 1966.

Nutting, Wallace. *Furniture Treasury.* New York: The Macmillan Company, 1928.

Oman, C.C. "An XVIIIth Century Record of Silver Furniture at Windsor Castle" *The Connoisseur,* volume XCIV, July—December, 1934.

Praz, Mario. *An Illustrated History of Interior Decoration, from Pompeii to Art Nouveau.* New York: Thames and Hudson, Inc., 1982.

Prime, Alfred Coxe. "John Elliott" Cabinet and Looking-Glass Maker of Philadelphia. *The Pennsylvania Museum Bulletin.* Philadelphia: The Pennsylvania Museum and School of Industrial Art, Volume XIX, No. 85. April 1924, pages 127-138.

Ring, Betty. "Checklist of looking-glass and frame makers and merchants known by their labels," *The Magazine Antiques,* May, 1981, page 1178.

Schwarz, Robert D. *The Stephen Girard Collection, A Selective Catalog.* Philadelphia: Girard College, 1980.

Shepherd, Raymond V., Junior, "Cliveden and Its Philadelphia-Chippendale Furniture: A Documented History;" *The American Art Journal,* volume VIII, number 2. November, 1976.

Sheraton, Thomas. *The Cabinet-Maker and Upholsterer's Drawing-Book.* New York, Washington, and London: Praeger Publishers, 1970.

Schiffer, Margaret Berwind. *Furniture and Its Makers of Chester County, Pennsylvania.* Exton: Schiffer Publishing Limited, 1978.

Schweig, Dr. Bruno. *Mirrors, A Guide to the Manufacture of Mirrors and Reflecting Surfaces.* London: Pelham Books, 1973.

Strickland, Peter, "Documented Philadelphia looking-glass manufacturers, circa 1800-1850." *The Magazine Antiques,* April, 1976, page 794.

Symonds, R.W. *Furniture Making in Seventeenth and Eighteenth Century England, an outline for collectors.* London: The Connoisseur, 1955.

Wainwright, Nicholas. *Grandure in Colonial Philadelphia: The House and Furniture of John Cadwalader.*

Ward-Jackson, Peter. *English Furniture Designs of the Eighteenth Century.* Victoria and Albert Museum. London: Her Majesty's Stationery Office, 1959.

Wick, Wendy C. *Stephen Girard: A Patron of the Philadelphia Furniture Trade.* M.S. Thesis, University of Delaware, 1977. Pages 90, 105, 284 (illustrated).

Wills, Geoffrey. *English Looking Glasses: a study of the glass, frames and makers, 1670-1820.* London: County Life Limited, 1965.